Aqueous Lubrication

Natural and Biomimetic Approaches

IISc Research Monographs Series

World Scientific Publishing Company (WSPC), Singapore and Indian Institute of Science (IISc), Bangalore co-publish a series of state-of-the-art monographs written by experts in specific areas. They include, but are not limited to, the authors' own research work.

This pioneering collaboration aims to contribute significantly in disseminating current Indian scientific understanding worldwide. In addition, the collaboration also proposes to bring the best scientific thoughts and ideas across the world in areas of priority to India through specially designed India editions.

Series Editor:

K. Kesava Rao (kesava@chemeng.iisc.ernet.in)

Editorial Board:

H. R. Krishnamurthy (hrkrish@physics.iisc.ernet.in)
P. Vijay Kumar (vijay@ece.iisc.ernet.in)
Gadadhar Misra (gm@math.iisc.ernet.in)
S. Ramasesha (ramasesh@sscu.iisc.ernet.in)
Usha Vijayaraghavan (uvr@mcbl.iisc.ernet.in)

Books in the Series:

Vol. 1   V. R. Voller, *Basic Control Volume Finite Element Methods for Fluids and Solids* (2009).

Vol. 2   Deepak D'Souza and Priti Shankar (Eds.), *Modern Applications of Automata Theory* (2012).

Vol. 3   Nicholas Spencer (Ed.), *Aqueous Lubrication* (2014).

IISc Research Monographs Series

Aqueous Lubrication

Natural and Biomimetic Approaches

Nicholas D. Spencer

ETH Zurich, Switzerland

**IISc
Press**

W♭ World Scientific

NEW JERSEY · LONDON · SINGAPORE · BEIJING · SHANGHAI · HONG KONG · TAIPEI · CHENNAI

Published by

World Scientific Publishing Co. Pte. Ltd.

5 Toh Tuck Link, Singapore 596224

USA office: 27 Warren Street, Suite 401-402, Hackensack, NJ 07601

UK office: 57 Shelton Street, Covent Garden, London WC2H 9HE

British Library Cataloguing-in-Publication Data
A catalogue record for this book is available from the British Library.

IISc Research Monographs Series — Vol. 3
AQUEOUS LUBRICATION
Natural and Biomimetic Approaches

Copyright © 2014 by World Scientific Publishing Co. Pte. Ltd.

ISBN 978-981-4313-76-6

In-house Editor: Rhaimie Wahap

Typeset by Stallion Press
Email: enquiries@stallionpress.com

Preface

Water, in the presence of a number of additives, meets all nature's lubrication needs. Mankind, however, has largely used oil for lubrication, mainly because of oil's higher viscosity and pressure-viscosity coefficient, but also for reasons connected with the way in which lubricants are applied. These include issues related to volatility and corrosion. There remains, however, considerable scope for the application of water lubrication in machines that could exploit its advantages over oil, such as greater environmental compatibility, better heat-transfer properties, and the potential for ultralow friction.

This book is about aqueous lubrication in nature, and man's attempts to understand it and imitate it. It deals with natural lubrication mechanisms as found in articular joints (Crockett, Chap. 1) and in mucous membranes (Lee, Chap. 2). It also describes human efforts to interface with natural lubrication systems in the context of food (Stokes, Chap. 3) and personal-care products (Luengo *et al.*, Chap. 4). Finally, it covers the current state of our understanding of the principles of aqueous lubrication with gels (Liu and Gong, Chap. 5) and with polymer brushes (Giasson and Spencer, Chap. 6), as well as efforts to apply aqueous lubrication to materials (Martin and De Barros-Bouchet, Chap. 7, and Kalin, Chap. 8).

The idea for this book was born during my period as Centenary Visiting Professor at the Indian Institute of Science, Bangalore. I am deeply indebted to Professor Sanjay K. Biswas and the Director of the Institute, Professor P. Balaram, for making my stay possible, thereby creating a lasting connection to their institution and to India. Very sadly, Professor Biswas passed away suddenly in April 2013, during the final stages of the preparation of the book. He will be deeply missed by the many people whose lives he touched, and I would like to dedicate this book to his memory.

Nicholas D. Spencer
Laboratory for Surface Science and Technology
Department of Materials, ETH Zurich, Switzerland
nspencer@ethz.ch

Contents

Preface   v

Chapter 1.   Tribology of Natural Articular Joints   1
   Rowena Crockett

   1.1   Articular Joints   1
   1.2   The Structure of Natural Articular Cartilage   3
   1.3   The Surface of Cartilage   5
   1.4   Friction and Wear of Natural Cartilage   10
   1.5   Lubrication Studies with Artificial Surfaces   13
   1.6   Lubrication Theories   16
   1.7   Conclusions   23
   References   25

Chapter 2.   Sticky and Slippery: Interfacial Forces of Mucin
                and Mucus Gels   33
   Seunghwan Lee

   2.1   Introduction   33
   2.2   Molecular Mucins   35
   2.3   Mucus and Mucosa   51
   2.4   Summary and Outlook   64
   2.5   Acknowledgements   66
   References   66

Chapter 3.   Aqueous Lubrication and Food Emulsions   73
   Jason R. Stokes

   3.1   Introduction   73
   3.2   Emulsion Lubrication in Engineering   74

3.3   Emulsion Lubrication in Soft-Tribology and Food
      Applications   84
3.4   Outlook for Emulsion and Food Emulsion Lubrication . . .   98
Acknowledgements   99
References   99

Chapter 4.   Aqueous Lubrication in Cosmetics   103
Gustavo S. Luengo, Anthony Galliano and Claude Dubief

4.1   Introduction. The importance of aqueous lubrication
      in Cosmetic Science 103
4.2   The Cosmetic Substrate 104
4.3   The Effect of water on hair structure 109
4.4   Cosmetic Tribology. Lubrication Mechanism 110
4.5   Lubrication evaluation 118
4.6   Hair Care Products: Ingredients and Formulation 126
References 140

Chapter 5.   Hydrogel Friction and Lubrication   145
Jian Liu and Jian Ping Gong

5.1   Introduction 145
5.2   Experimental Details 147
5.3   A Model of Gel Friction: Repulsion and Adsorption 150
5.4   Frictional Properties of a Neutral Hydrogel: PVA Gel . . . 153
5.5   Frictional Properties of Polyelectrolyte Hydrogels 160
5.6   Friction of Hydrogels with Surface-Modified Structure . . . 170
5.7   Application of Robust Gels with Low Friction as
      Substitutes for Biological Tissues 174
5.8   Summary 178
References 179

Chapter 6.   Aqueous Lubrication with Polymer Brushes   183
Suzanne Giasson and Nicholas D. Spencer

6.1   Introduction 183
6.2   Fundamental Aspects of Lubricating
      with Polymer Brushes 194
6.3   Macro-, Micro- and Nano-Tribological Measurement
      Approaches for Polymer Brushes 196
6.4   Experimental Studies of Neutral and Charged Systems . . 198

6.5 Conclusion 214

References 214

Chapter 7. Water-Like Lubrication of Hard Contacts
           by Polyhydric Alcohols   219
Jean Michel Martin and Maria Isabel De Barros-Bouchet

7.1 Introduction 219

7.2 Polyhydric Alcohols and Carbohydrates as Lubricants . . . 221

7.3 Lubrication of Steel by Glycerol 223

7.4 Lubrication of Diamondlike Carbon by Glycerol 230

7.5 Evidence of Water Formation by Computer Simulation . .   231

References 235

Chapter 8. Aqueous Lubrication of Ceramics   237
Mitjan Kalin

8.0 Introduction 237

8.1 Oil vs. Water Lubrication Technology 238

8.2 Super-Low Friction of Non-Oxide Ceramics 241

8.3 Wear-Protective Hydrated Tribochemical Layers 244

8.4 The Electrochemical Mechanism of pH and the Electric
     Charge at the Surfaces in Water 248

8.5 Effects of pH and Surface Charge on Tribological
     Behaviour and Formation of Boundary Surface Layers
     of Oxide Ceramics 250

8.6 Comments on Various Influencing Parameters 259

8.7 Conclusions 266

References 267

Index   269

Chapter 1

Tribology of Natural Articular Joints

Rowena Crockett

Swiss Federal Laboratories for Materials Science and Technology Empa
Ueberlandstrasse 129, 8600 Duebendorf, Switzerland
Rowena.Crockett@empa.ch

1.1. Articular Joints

Articular or hyaline cartilage is a smooth, white tissue that covers the
surface of all moving joints in the human body. The most abundant sub-
stance in cartilage is water, which makes up 70 to 80% of the weight.[1,2]
The remaining material includes type II collagen, which forms the fibrous
network, and the proteoglycan aggregate, a large complex of aggrecan and
hyaluronan. The mechanical properties of cartilage are attributed to the
highly hydrated aggregate trapped in a matrix of collagen.[3,4] Smaller proteoglycans and chondrocytes are also present in the extra-cellular matrix
of articular cartilage.[5] Collagen makes up approximately 50–75% of the
dry weight of articular cartilage, whereas the proteoglycans account for
approximately 15–30%.[2,5]

Articular joints are lubricated by synovial fluid — a viscous, non-
Newtonian liquid. Most of the proteins in synovial fluid are from blood
serum that has been dialysed through the synovial membrane, known as
the *synovium* (Fig. 1.1).[6] The *synovium* is a thin, soft tissue inside the joint
capsule that is made up of two layers (Fig. 1.1). The inner layer is called
the *intima* and contains fibroblast-like cells called synoviocytes, which provide hyaluronan and lubricin, amongst other proteins, to the synovial fluid.
Additional glycoproteins are produced by the cartilage and *synovium* itself.
In addition, synovial fluid contains the glycosaminoglycan hyaluronan, the
deprotonated form of hyaluronic acid. The concentration of hyaluronan in
synovial fluid is, at 3–4 mg/mL, the highest in the human body.[7]

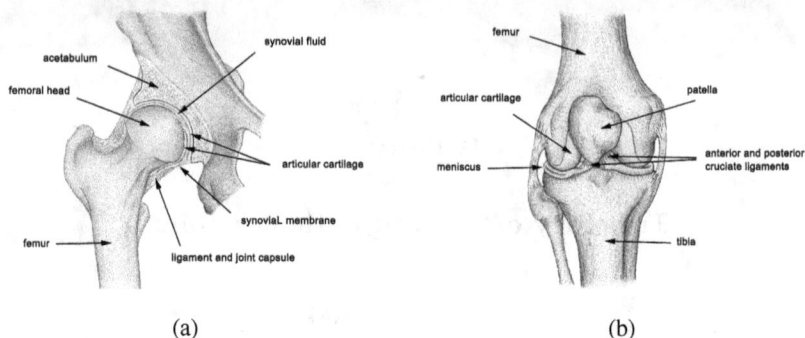

Fig. 1.1. Schematic diagrams of (a) natural hip and (b) natural knee joints (diagrams were kindly provided by Sasa Vranjkovic, Empa, reproduced from the *Encyclopedia of Tribology* with kind permission from Springer Verlag).

The remarkable mechanical properties and low friction and wear of cartilage have been of interest to researchers for over 200 years. Earlier studies focused more closely on the chemical nature of cartilage, for example on the extraction and identification of chondroitin sulphate.[8–10] It was noted as early as 1742 by William Hunter that cartilage is a non-vascular material and, therefore, most early interest was directed towards the ability of cartilage to repair itself and obtain nutrition.[11,12] Initially, investigations into the lubrication mechanism of articular cartilage led to the conclusion that a thick layer of synovial fluid was maintained between the surfaces during motion. It was proposed that the fluid flowed between the surfaces according to the laws of viscous hydrodynamics and that a weeping mechanism squeezed the fluid into the joint when pressure was applied.[13,14] Since that time, a great deal of progress has been made in the field of interstitial fluid pressurization and in understanding its role in the lubrication of cartilage.[15] It was also proposed that some contact between the surfaces took place during motion.[16]

While progress was quickly made in determining the biochemical composition and structure of the bulk of cartilage, studies on the chemical nature of the surface and the influence of components of synovial fluid on lubrication have provided a number of seemingly contradictory theories. Many of the difficulties in describing the friction behaviour of biomolecules at the surface seem to arise from the apparently confusing terminology used for the structure of cartilage as well as some of the technical drawbacks inherent in early analytical methods.

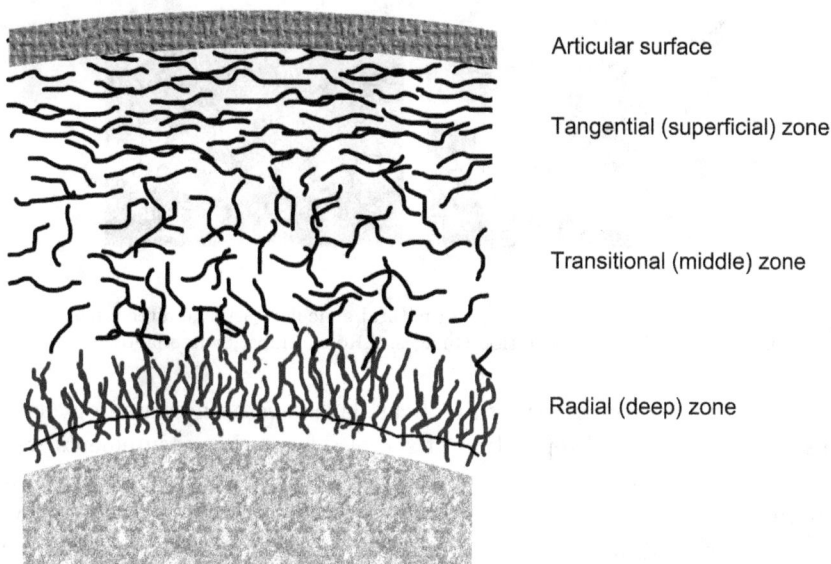

Articular surface

Tangential (superficial) zone

Transitional (middle) zone

Radial (deep) zone

Fig. 1.2. The bulk structure and surface of articular cartilage.

A number of reviews have been published covering the various aspects of cartilage lubrication and the differing theories that have been proposed.[2,3,16−20]

1.2. The Structure of Natural Articular Cartilage

The bulk of cartilage is split into four zones that are distinguished according to the orientation of collagen and concentration of proteoglycan (Fig. 1.2).[3,21,22] The uppermost zone of the bulk of cartilage is known as the tangential or superficial zone and is made up of a relatively large quantity of collagen that is arranged parallel to the surface with a relatively low proportion of proteoglycan.[23] One of the common sources of confusion is that this "superficial" zone is *below* the articular surface (see below) and it will, therefore, be referred to as the tangential zone in this chapter (Fig. 1.2). In the transitional or middle zone, there is a random arrangement of collagen and the largest proportion of proteoglycans. In the deep zone, the collagen fibres are oriented perpendicular to the surface, and the partly mineralised calcified zone provides the transition from bone to cartilage. The number, shape and orientation of the cells, chondrocytes, also vary throughout the cartilage.[21,22]

Fig. 1.3. Environmental SEM images of natural articular cartilage after washing with (a) PBS and (b) 1.5 M NaCl solution (Images: S.Roos, Empa, Switzerland).

Many studies of the structure of cartilage have been carried out using electron microscopy. Preparation of samples for such measurements were often harsh and included washing the samples with solutions of high salt concentrations or mechanical scrubbing of the surface.[21,24] Figure 1.3 shows the effect of solutions with low and high salt concentrations on the appearance of articular cartilage.

These severe cleaning techniques led to the observation that the articular surface of cartilage was the collagenous tangential zone. It was thus concluded that the collagen matrix was not separated by any structure from the synovial fluid and the opposing cartilage.[21] As any proteoglycans at the surface would certainly be leached out into the synovial fluid, such a model implies that the system consists of collagen sliding against collagen.[25] The adsorption of a homogeneous layer of biomolecules from the synovial fluid onto the open collagen network would, however, be highly unlikely. The presence of a non-collagenous articular surface covering the tangential zone has been observed by a number of research groups.[26–30] As well as being an articular surface, this layer has also been called the *lamina splendens*,[3] the superior surface and the surface amorphous layer,[27,28,31] although confusingly the term surface amorphous layer is sometimes also used to describe the tangential (superficial) zone after exposure to air.[21] The name "superficial zone" and its common confusion with the "surface amorphous layer" seems to have led some researchers to believe that bulk proteoglycans such as aggrecan and biglycan are at the articular surface and may, therefore, participate in the lubrication mechanism. As the tangential zone lies beneath the articular surface, it will undoubtedly influence the mechanical and, therefore, the tribological properties of the system but cannot be involved in the mechanism of low friction resulting from the interaction of synovial fluid with the surface.[15,32–34] Immediately after removal from the joint for experimental studies, another layer is present on top of the articular surface, which is sometimes also referred to as the superficial layer, surface amorphous layer or *lamina splendens*, although in these cases, the articular surface is generally absent from the proposed description.[30,35,36] Meachim *et al.* described a superficial layer that seems to consist of both the articular surface and the layer that is adsorbed or deposited onto it from the synovial fluid.[37] In this chapter, the layer directly above the tangential zone will be referred to as the articular surface, as shown in Fig. 1.2, as this cannot be easily removed from the cartilage The synovial-fluid-derived superficial layer on top of the articular surface can be removed by washing thoroughly with PBS.

1.3. The Surface of Cartilage

One of the earliest research problems associated with the articular surface was an inability to find a consistent and accepted description of its appearance. Gardner, in 1971, blamed this on a small army of students with varying aptitudes for drawing diagrams, and also on the wide variety of methods used to observe the surface.[38] There was, and is, a tendency to assume that most other researchers were simply mistaken. However, if one were to assume that the artistic and experimental skills of the students were satisfactory then the only other possible source of the deviations observed in the literature is in the material itself.[27] The transition in the appearance of the surface from smooth and featureless to the rough and fibrous layer sometimes described as *lamina splendens* may be attributed to the inevitable loss of water, the major component of articular cartilage,[3] following explantation. In much the same way, a raisin does not have the same appearance as a grape.

Among the large number of terms that have been coined to denote the surface is the seemingly excessive "uppermost superficial surface layer", which imparts the impression that cartilage is in a position to form further surfaces *ad infinitum*.[29] However, such phrases have become necessary, since it is often not obvious which "surface" is being referred to in the literature, making it difficult to compare experimental results. If the sample is not allowed to dry, for example, then the *lamina splendens* will not be observed at all. If the sample is washed thoroughly with phosphate buffered saline (PBS) then the layer adsorbed or deposited onto the articular surface will not be observed (Fig. 1.4(a)). Many studies concerning the surface of articular cartilage have reported that it is hydrophobic immediately after removal from the joint. However, without the adsorbed/deposited layer, the surface appears to be neither smooth nor hydrophobic.[30] A number of reports raise the issue that removal of cartilage from the joint and sample preparation will affect the appearance of the sample.[27,39,40]

Atomic force microscopy imaging allows the articular surface to be studied in an aqueous environment and was first carried out on this material by Jurvelin *et al.* using relatively mild sample-preparation conditions.[35] The images of the hydrated articular surface showed that it was an amorphous, non-collagenous layer. None of the surface features that had previously been observed with SEM were seen in the AFM images. Removal of the articular surface by digestion with chondroitinase AC revealed the collagenous tangential zone, which could be easily distinguished from the articular surface layer.[35]

A comparison of the images of hydrated cartilage acquired with environmental scanning electron microscopy (ESEM) and samples prepared for SEM carried out by Forster *et al.* demonstrate the difference caused by the preparation technique.[40] Not only did the sample preparation for SEM lead to a change in the appearance of the articular surface, denoted as the surface lamina, but it also removed or damaged parts of the articular surface. In this study, a further layer was observed on top of the articular surface and called the boundary layer.[40] A similar region was also observed on cartilage in an aqueous environment using AFM. When the excised cartilage was imaged in PBS immediately after removal, a surface gel layer was observed, however, removal of this with chondroitinase ABC revealed the collagenous tangential zone and it was, therefore, not distinguished from the articular surface.[41] The uppermost superficial surface layer, described by Kumar *et al.* was also found to be highly viscous and non-fibrous.[29] Removal of this by digestion with alkaline protease again exposed the fibrous tangential zone. However, the descriptions given for this layer suggest that it is the one on top of the articular surface, as observed by Forster *et al.*[29,40] If chondroitinase and alkaline protease digest the articular surface, then anything on top of this will undoubtedly also be dispersed.

We have studied the various layers that have been denoted as the surface by means of ESEM. Figure 1.4 shows the effect of dehydration on the layer denoted as the articular surface in Fig. 1.3 during ESEM imaging, as well as the effect on the layer that is found on top of this immediately after removal from the joint.[30] It can be seen in both cases that loss of water has a very dramatic effect and that both articular surface and the layer on top of it appear to be fibrous. It is possible that both layers become so thin

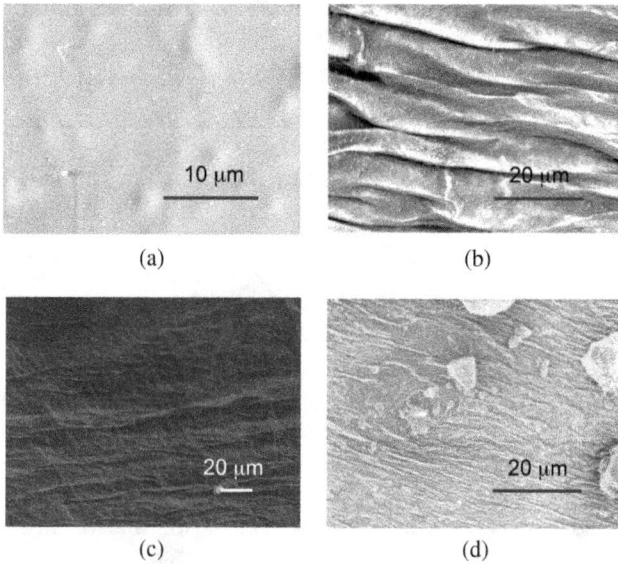

(a)   (b)

(c)   (d)

Fig. 1.4.   ESEM images of articular cartilage (a) immediately after removal from syn-ovial fluid and (b) after 15 minutes in the ESEM. This imaged layer is not shown in Fig. 1.2 as it can easily be removed with PBS and is therefore not an integral part of the cartilage. (c) The articular surface shown in Fig. 1.3(a) after drying out and (d) the tangential zone visible after the articular surface, as shown in Fig. 1.2, has been removed with a scalpel. Reproduced from reference 30 with kind permission.

on drying out that the collagen of the tangential zone is visible through the layers. This would suggest that both layers above the tangential zone contain a large amount of water. For comparison, the tangential zone is also shown in Fig. 1.4.[30]

The apparent hydrophobic nature of cartilage immediately after removal from the natural environment can also be attributed to dehydra-tion. Cartilage from the knee joint of a cow was immersed, immediately after removal, in PBS and changes at the surface were followed by measuring force-distance curves with AFM (Fig. 1.5).[30] Initially, the adhesion was very high between the AFM silicon nitride tip, which tends to be hydrophobic due to contamination, and the sample surface, but it decreased with time, presumably due to rehydration.

This initial high adhesion was attributed to strong hydrophobic interac-tions. As time passed, the indentation depth of the tip increased for constant applied load. It was concluded that the upper layer, denoted the superficial gel-like layer (Fig. 1.5(a)), not only became softer through the absorption of

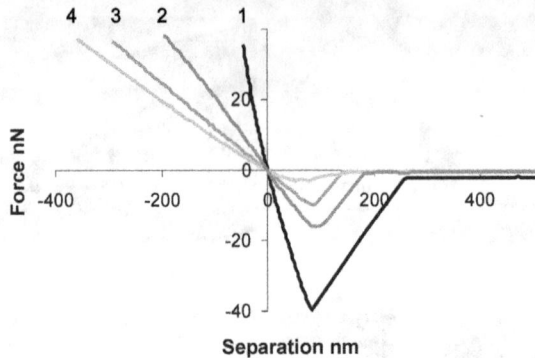

Fig. 1.5.   Force-distance curves measured on separation of the AFM tip from the surface at (1) 20 min., (2) 22 min., (3) 24 min., and (4) 26 min. after submersion of the bovine cartilage in PBS. Reproduced from Reference 30 with kind permission.

water but also underwent a chemical reorganisation that resulted in the loss of hydrophobicity.[30] Therefore, it was proposed that after removal from the synovial fluid, the evaporation of water from the gel-like layer was accompanied by the diffusion of hydrophobic molecules to reduce the surface energy. Removal of the gel-like layer either by washing thoroughly with PBS or by mechanical removal from a frozen sample exposed the articular surface (Fig. 1.5(c)). This region had a granular appearance in the AFM image and did not show the highly viscous behaviour of the superficial gel-like layer. No adhesion was detected between the articular surface and the AFM tip when immersed in PBS, as had been observed for the superficial gel-like layer, and no change in the force-distance curves was observed over a period of 1 hour.[30] When immersed in synovial fluid, neither the articular surface nor the gel-like layer showed any changes within 1 hour.

To summarise, differences in appearance, mechanical properties, or thicknesses of surface layers reported in the literature could all be attributed to both the loss of water from the sample and the possibility that completely different surfaces are being examined.

While many of the differences in topographical assessments of what has in some way been denoted as being a surface can be attributed to varying degrees of hydration, the detection of biochemical differences clearly indicates that in some cases, truly different surface are being described. One of the earliest biochemical investigations into the articular surface indicated that this was made up of two layers.[42] These layers were distinguished using a staining technique with a cationic dye, cupromeronic blue, and digestion with hyaluronidase. It was found that the upper layer was hyaluronidase-resistant and the sulphated glycosaminoglycans chondroitin sulphate and keratin sulphate were not detected.[42] This upper layer of the articular surface was also found to be negatively charged, an observation later confirmed using cationic ferritin.[43] The lower, thicker layer in the articular surface was not resistant to hyaluronidase digestion and did contain sulphated glycosaminoglycans.[42] It was concluded that a lower layer of proteoglycan provides a continuum between the collagenous tangential zone and a thin layer of glycoprotein. A more detailed analyses of the whole articular surface, referred to as the surface amorphous layer, was carried out by Graindorge *et al.* Large amounts of lipids, protein and sulphated glycosaminoglycans were found but no hydroxyproline was detected, confirming the observation from ESEM and AFM images that the articular surface does not contain collagen. However, contrary to the investigations of Orford and Gardner, hyaluronan was not detected.[42] On the other hand, Balazs *et al.* who found two surface layers showed that the upper layer was non-collagenous and made up mostly of hyaluronan. The second non-collagenous layer was also found to contain non-sulphated glycosaminoglycan.[44] It is clear from the large variations in the observed chemistry at the surface of cartilage that there are a number of distinguishable collagen-free layers above the tangential zone.

A number of investigations have been carried out using AFM to follow the effect of enzyme digestion on the observed surface with the aim of gaining information about its chemical nature.[29,35,41] For example, chondroitinase treatment, in one study, removed the upper layer exposing the collagenous tangential zone.[41] In another study, chondroitinase only partially removed the imaged layer whereas alkaline protease completely removed it and hyaluronidase had no effect.[29,35] It would appear that different layers were being investigated in these studies. However, it should be noted that the digestion of a lower layer could affect the integrity of an upper layer if the layers chemically interact with each other.

As already stated, immediately after removal of the cartilage from the joint, the surface of the sample appears smooth and is hydrophobic. We investigated the chemical nature of this hydrophobicity by pressing a clean silicon wafer onto the sample and measuring the infrared spectrum of the thin layer that adhered to the silicon with attenuated total reflection infrared spectroscopy (ATR-IR).[36] We found that this was made up mostly of phospholipids. However, as was shown with AFM, the cartilage only becomes hydrophobic after it is removed from the synovial fluid or aqueous solution.[30] The whole upper, or superficial, layer can be relatively easily removed and therefore appears to be deposited or adsorbed on the surface rather than being an integral part of the cartilage. It could be shown with IR that this layer is not simply synovial fluid, as the relative amounts of hyaluronan and protein are very different. This superficial layer was found to consist mostly of hyaluronan.[36]

Although it is not entirely clear what the biochemical composition of the articular surface is, the various investigations that have been reported do allow a number of possibilities to be ruled out. Firstly, at whatever position the surface is above the tangential zone, it is non-collagenous.[29,30,35,40,44] Additionally, the lack of sulphated glycosamino-glycans indicates that there is no or very little aggrecan or proteoglycan aggregate present.[42,44]

1.4. Friction and Wear of Natural Cartilage

In many tribological studies on artificial materials, wear generally results in a new surface being produced that is chemically the same or at least very similar to that which was removed. Additionally, the surface is usually chemically well defined and characterized, and there is no ambiguity concerning its position in the tribological system. None of these conditions apply when the material being studied is natural cartilage. Not surprisingly, the variation in the results of friction and wear tests reflects the number of possible surfaces that are being subjected to tests.

The superficial layer, at the very top of the cartilage sample after removal from the joint, is washed off in PBS (Fig. 1.6). Therefore, in those cases where the cartilage samples used in friction and wear studies were stored in or washed with aqueous buffer or salt solutions, this layer is no longer present when the samples are measured. Studies have, however, been carried out without pre-treatment on intact joints in the presence of the superficial layer. Measurements on the cartilage of intact temporomandibular joints from swine found friction coefficients of 0.0164 ± 0.0020. The friction increased to 0.0223 ± 0.0050 after washing with PBS, and further to 0.0398 ± 0.0047 after scouring the surface with gauze.[24] SEM images of the samples indicated that the articular surface was intact after washing with PBS but severely damaged by the gauze scouring.[24] A similar behaviour was determined with eleven canine hip joints where the friction coefficients increased from 0.002 to 0.010 in the intact joints to 0.008 to 0.027 after washing with saline solution.[45] Friction coefficients of 0.0089 ± 0.0010 were

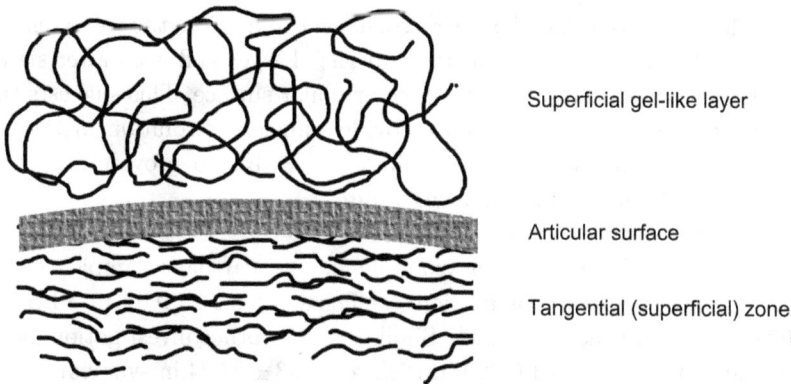

Superficial gel-like layer

Articular surface

Tangential (superficial) zone

Fig. 1.6. The layers of cartilage that have been investigated, and referred to as being the surface. The uppermost layer has been referred to as the boundary layer and the superficial gel-like layer.[30,40,41] Based on experimental evidence, Orford proposed that the articular surface was made up of two layers.[42] In some cases, the articular surface and the superficial gel-like layer have been investigated as one layer, referred to as the superior surface and the surface amorphous layer.[27,28,31,37] Based on experimental evidence, Orford proposed that the articular surface was made up of two layers.[42] Additionally the tangential zone has been referred to as the amorphous superficial layer.[21] However, a number of studies have shown that the articular surface is non-collagenous.[26−30,40]

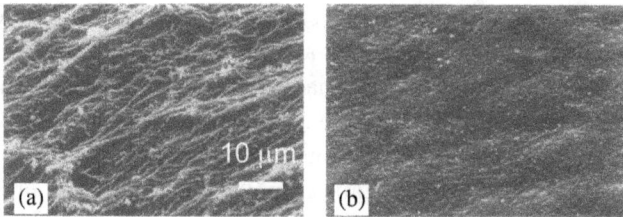

Fig. 1.7. The surface of cartilage from the hind limb of a rabbit after treating with papain for 4 days (a) compared to cartilage from an untreated joint (b). An increase in the coefficient of friction was observed on removing the articular surface using papain digestion but the value was still very low at 0.0131.[46] This suggests that the friction coefficient cannot be used as a reliable indication of the condition of the cartilage. Reproduced from Reference 46 with kind permission from *Clinical Biomechanics* (Elsevier).

found for intact rabbit stifles, which increased to 0.0131 ± 0.0024 following damage induced to the articular surface by papain digestion.[46] (Fig. 1.7).

The friction coefficient of intact canine joints, lubricated with bovine synovial fluid was found to be 0.0035 to 0.0040 at 37°C.[47] In this case, the samples were measured first in buffer solution and the superficial layer had, therefore, probably been removed. These measurements indicate that while the intact joint has the lowest coefficient of friction, the value remains relatively low after washing off the superficial layer and even after significant damage has been inflicted on the articular surface. This suggests that friction measurements might not provide sufficient resolution in the biochemical environment of natural joints to be able to provide a detailed insight into the natural mechanism of lubrication.

Friction coefficients tend to be higher when the cartilage samples are removed from the joints. For example bovine cartilage samples sliding against each other have a friction coefficient of 0.24 ± 0.04 in PBS and 0.028 ± 0.006 in synovial fluid.[48] Similarly, in another investigation values have ranged from 0.185 ± 0.020 in PBS to 0.033 ± 0.004 in synovial fluid.[49] In both of these studies, the superficial layer was removed prior to the friction tests by soaking in PBS. The relatively high friction coefficients for cartilage in synovial fluid in comparison to those found for the intact joints suggest that the articular surface may also have been damaged. It is, however, possible that some other change occurs to the system that has a negative effect on the tribological system. One clear difference between the *in vivo* joint and the samples of excised cartilage is that the former is not immersed in synovial fluid, as the quantity in for example a knee joint is only between 0.5 and 4.0 mL.[50] This is entrained into the articular joint during motion, whereas excised cartilage is generally immersed in fluid during sliding experiments. As with any biological material, removal from the living system will undoubtedly have an effect on cartilage but to what extent changes to the bulk cartilage or the articular surface influence the tribological system is unclear. Friction coefficients as low as 0.04 have been measured for cartilage sliding against cartilage in Ringer's solution[31] — a value significantly lower than that found for sliding in PBS. This difference suggests that either different layers are exposed and become the surface or there are large differences within the bulk cartilage. The value measured in Ringer's solution was not significantly different when the articular surface was removed with SDS to expose the collagenous tangential zone.[31] As the articular surface is very thin, only a small amount of wear during sliding will result in significant damage to or complete removal of this layer resulting in a very different tribological system than that found in the living body, regardless of whether the friction coefficient increases, decreases, or stays the same. There is no obvious reason to assume that interactions of the non-collagenous articular surface with biomolecules found in synovial fluid will be the same or even similar to those of the collagenous tangential zone.

It has been observed that the addition of hyaluronan to PBS decreases the friction coefficient of cartilage sliding against cartilage from 0.29 ± 0.01 in PBS to 0.12 ± 0.01, with a similar decrease to 0.11 ± 0.01 being measured when the proteoglycan PRG4 is added to PBS. Solutions containing both hyaluronan and PRG4 resulted in a further decrease to 0.066 ± 0.003. However, the friction coefficient in synovial fluid at 0.025 ± 0.005 suggests that the articular surface was no longer intact when compared to the values measured for the intact joints.[51]

One of the difficulties encountered when measuring cartilage sliding against cartilage is achieving a well-defined contact area. This is sometimes overcome by using steel or glass surfaces as the counter-face sliding against cartilage, thus removing the measured system further from that of the natural articular joint.[32,33,40,52−57] The friction coefficients measured for such systems tend to be higher than those measured for cartilage sliding against cartilage, rising to values as high as 0.5 in some reported cases.[40]

A fundamental question that has not been answered by the many tribological measurements on cartilage is whether or not the articular cartilage/synovial fluid system is capable of functioning outside the body in the same or even a similar way to that of the *in vivo* joint. Healthy, natural joints *in vivo* are never exposed to buffer or Ringers solution. It is, therefore, possible that such solutions induce a degree of damage that cannot be rectified simply by submersing the sample in synovial fluid.

1.5. Lubrication Studies with Artificial Surfaces

Natural articular cartilage sliding in synovial fluid is an exceptionally complicated system. A vast number of proteins, glycoproteins and other biomolecules, such as hyaluronan and phospholipids, are present. Added to this are the uncertainties as to where the articular surface is and what it consists of. As a result, artificial systems have been used to provide more controllable and well-defined surfaces for investigations into components of synovial fluid that may be responsible for low friction. Such systems no longer take into account the influence of the mechanical properties and high water content of cartilage, however, they do allow the ability of biomolecules to reduce friction to be individually assessed.

Potential lubricating mechanisms have been investigated on the molecular level with the surface forces apparatus (SFA), which allows normal and shear forces to be measured to a high degree of accuracy.[58−66] Biomolecules that have been investigated with this technique include

hyaluronan, lubricin, as well as complexes of hyaluronan and aggrecan. Negatively charged mica surfaces are generally used in SFA experiments, and
the large, negatively charged polysaccharide hyaluronan does not adsorb
strongly to this.[61] At high loads, hyaluronan was also squeezed out of the
contact between surfaces that had been modified with positively charged
surfactant bilayers and also when it was adsorbed onto the surface via
positively charged calcium ions.[60,62] It was concluded from this that if
hyaluronan was to act as a boundary lubricant in natural joints it must
be chemically or specifically bound to the surface.[60]

The proteoglycan lubricin was adsorbed onto bare mica as well as chemically modified surfaces to form layers of 60–100 nm thickness.[64] At low
pressures, low friction coefficients of 0.02–0.04 were found for lubricin on
the negatively charged mica, as well as for surfaces that had been rendered positively charged with poly L-lysine or aminothiol. At pressures
above about 600 kPa, the adsorbed lubricin rearranged under shear and
the friction coefficient increased to values above 0.2. When the lubricin
was adsorbed onto the hydrophobic surface of gold coated with alkanethiol,
the friction coefficient was higher than 0.3 at low pressures.[64] There was
a large range of values for the friction coefficient on both the hydrophobic
and hydrophilic surfaces, suggesting that the lubricin can adsorb in a wide
variety of different conformations. Further studies showed that the lubricin
aggregates by forming disulphide bonds between the ends of the glycoproteins. Following removal of these bonds by reduction and alkylation of the
cysteine residues, the lubricin was still able to adsorb onto the mica surface
but gave a higher friction coefficient of 0.13 to 0.17 in the SFA at low loads.
Both the native form and the reduced lubricin were found to protect the
mica surfaces from wear.[63]

Aggrecan is a large proteoglycan with a molecular weight of 1 to 3 MDa,
which attaches to hyaluronan at hyaluronan binding sites in the G1 domain
at the N-terminal region. This interaction is stabilized by link proteins.[67]
The resulting complex, known as proteoglycan aggregate, is trapped in the
compressed form in the collagen network of bulk articular cartilage and
provides cartilage with resistance to compression.[68] If this aggregate were
small enough to pass through the articular surface then it would quickly
be depleted from the cartilage, resulting in catastrophic damage to the
joints. The frictional properties of proteoglycan aggregate have been investigated using a surface forces apparatus.[58] In these experiments, biotiny-
lated hyaluronan was first attached to avidin-coated mica and then the
hyaluronan-coated samples were incubated in a solution of aggrecan and

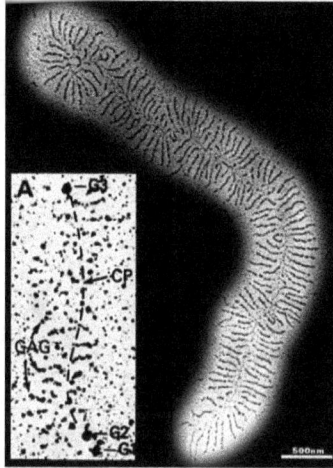

Fig. 1.8. Electron micrograph of the proteoglycan aggregate formed from hyaluronan, link protein and aggrecan (reproduced from Ref. 68 with kind permission from Glycoforum). Aggrecan in cartilage occupies less than 15% of its fully expanded volume in solution.[68]

link protein, which self-assemble to form the proteoglycan aggregate at the surface (Fig. 1.8). The resulting proteoglycan aggregate was found to have a very low friction coefficient.

Another model system that is often used to study the friction properties of synovial fluid or components of synovial fluid is latex sliding against glass.[34,69,70] This has often described as a system that can imitate cartilage sliding against cartilage and is not only used to study individual components found in the articular joint but also whole synovial fluid. The most abundant component of synovial fluid is albumin, which readily adsorbs onto hydrophobic and charged surfaces. Most investigations into the lubricating behaviour of albumin have found that this increases friction, or at least cannot lead to a significant decrease in the friction coefficient.[71] It can therefore be concluded that, unlike glass and latex, the articular surface is either protein resistant or only allows specific protein adsorption. This difference in the chemistry of the surface will lead to different adsorption behaviour of the many components of synovial fluid, which will not necessarily translate into a change in the friction coefficient, as this is not chemically specific.[72] The interactions or lack of them between the articular surface of cartilage and synovial fluid proteins has not yet been imitated in artificial systems used to study the tribology of synovial joints. This is a major drawback in measurements designed to determine the lubricating ability of whole synovial fluid. As with the mica used for surface-force measurements, the glass-against-latex system can only provide information on the lubricating ability of individual components or limited combinations of specific components from synovial fluid when adsorbed onto artificial surfaces.[73–75]

1.6. Lubrication Theories

1.6.1. *Fluid-film lubrication*

Initially, it was widely believed that the cartilage surfaces are separated by a fluid film during the swing phase of walking while at the commencement of motion, liquid is squeezed out of the cartilage to allow elastohydrodynamic lubrication.[18] Boundary lubrication was thought to occur when the "dry" cartilage surfaces came into contact.[16] The squeeze-film mechanism was also proposed to explain low friction at high loads.[76,77] This was later replaced by a theory of interstitial fluid pressurization. The experimental as well as theoretical evidence that the pressurization of the fluid in cartilage is sufficient to support most if not all the applied load has been reviewed by Ateshian.[78] This interstitial fluid pressurization is a fundamental aspect of the tribological system and one that has, to date, not been mimicked in studies using artificial model surfaces. It has been proposed that this mechanism operates during the normal operation of the joints, whereas boundary lubrication operates when the pressurization subsides and solid contact occurs between the opposing cartilage surfaces.[78]

The development of these theories has largely been based on comparisons with man-made systems where hard, solid surfaces are in contact and are lubricated by oil. At high speed and low load, the surfaces are separated and lubrication is in the hydrodynamic or elastohydrodynamic regimes. As speed decreases or load increases, lubrication goes through the mixed regime and eventually reaches the boundary regime, where the opposing surfaces are in contact. These changes in regime are accompanied by changes in the friction coefficient, which allows the conditions required to achieve each regime to be determined. A characteristic feature of this description of lubricating regimes is that interactions between the surface and the oil play no or very little role in determining the lubricating properties. At most, an additive will be adsorbed onto the surface from the oil in order to reduce friction in the boundary regime after the oil has been squeezed out.

In natural articular joints, the base of the lubricant is water and no evidence has been presented to date that indicates that cartilage can dry out sufficiently in the natural joint to allow "solid contact" between the surfaces. Water has a high affinity to the many hydrophilic biomolecules found in natural joints, whereas oil generally only weakly solvates its many additives. The extent to which water participates in the tribological system will depend on how tightly it is held within the hydration layers of the biomolecules. Surface-force measurements carried out by Raviv *et al.* demonstrated that very low friction coefficients of 0.0006 to 0.001 could be achieved in aqueous systems under conditions that would generally be described as being in the boundary regime:[79] Polyelectrolyte brushes attached to the surface remained hydrated under load and through a combination of osmotic effects and resistance to mutual interpenetration of the polymers were able to maintain low friction (see Chap. 6). Thus, in aqueous-based lubrication, a change in friction cannot automatically be associated with or be a requirement for a change in the lubricating regime. The ability of the system to hold water at the surface will depend both on the chemistry of the surface and those compounds that attach or adsorb onto it, which therefore, should not be squeezed out during loading. The interaction between the surface and the components of synovial fluid will play a decisive role in determining which compounds remain between the cartilage surfaces. Thus, it is possible that the lubricating regime occurring in natural joints remains constant, even though there may be changes in the relative contributions of pressurization of liquid within the cartilage versus liquid trapped at a highly hydrophilic surface.

1.6.2. *Hyaluronan*

Early studies on the lubrication of articular joints concentrated on the role of hyaluronan. This was mainly due to the observation that synovial fluid contains plasma proteins at a lower concentration than in the blood but large amounts of hyaluronan.[80] Additionally, it had been found that the viscosity of synovial fluid of patients with diseased joints was significantly reduced in comparison to the fluid found in healthy joints.[81] The concentration of hyaluronan in healthy joints is far larger than that which would be required to fill the entire volume of the fluid with the normal unconstrained volume of the molecules.[82] A hyaluronan molecule has an expanded random coil structure that requires molecules to begin to overlap at concentrations of $1\,\mathrm{mg}\,\mathrm{mL}^{-1}$ or higher.[7] Thus, synovial fluid can be viewed as an overcrowded solution where the hyaluronan chains interact and intertwine. All other biomolecules in synovial fluid are contained within this network and thus their movement will therefore be severely restricted. It was found that although the major plasma proteins found in synovial fluid do not form complexes or aggregates with hyaluronan, they do influence the viscosity of solutions containing hyaluronan.[83]

A thick layer has been observed at the surface of articular cartilage when this is not washed rigorously, as described previously, and this has been attributed to a mixture of protein and hyaluronan (Fig. 1.6).[44,82] The formation of a thick film on cartilage has also been observed in other studies.[30,36,84] The uppermost layer on articular cartilage is often removed with the justification that it is residual synovial fluid adhered to the surface. However, it could be shown with infrared spectroscopy that this layer has a very different chemical composition when compared to synovial fluid, as described above.[36] The infrared spectrum indicates that this layer is made up mostly of hyaluronan, consistent with the observation that the properties of the gel-like layer could be mimicked with hyaluronan concentrations of approximately $70 \, \mathrm{mg \, mL^{-1}}$. compared to the concentration in synovial fluid of about $3 \, \mathrm{mg \, mL^{-1}}$. The concentration of protein in synovial fluid is approximately 60% of the dry mass whereas in the superficial layer on cartilage, protein was found at concentrations of only $4.5 \pm 2.2\%$ of the dry mass.[36]

Filtration of synovial fluid through cartilage resulted in the formation of a stable film with an increased concentration of hyaluronan[85] (Fig. 1.9). It can be envisaged that this is how the gel-like layer with a high concentration

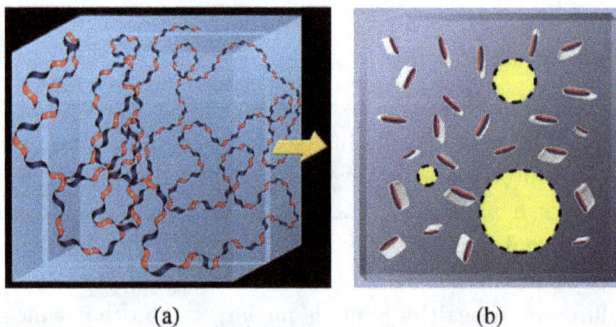

(a)   (b)

Fig. 1.9.   The structure of hyaluronan in aqueous solution. (a) The blue box represents the volume occupied by one molecule and the blue and red ribbon shows the alternating hydrophobic (red) and hydrophilic faces (blue) of the sugar residues. A volume of 1 mL is required by 1 mg of hyaluronan. (b) A slice through the box shown in (a) where the yellow circles indicate the space that is available to diffusing molecules (reproduced from Ref. 7 with kind permission from Glycoforum).

of hyaluronan forms at the surface under physiological conditions. This may also be the mechanism by which the cartilage obtains nutrients. The contacting surfaces of articular cartilage are not immersed in synovial fluid during rest, the fluid being entrained during articulation. The pressure applied to the cartilage during movement may squeeze out some of the water, which will then be re-adsorbed from the synovial fluid when loading is stopped. This process will result in an increase in the concentration of hyaluronan, as this is too large to pass through the articular surface, and a decrease in the concentration of proteins that are sufficiently small to pass through.

This process of concentrating hyaluronan at the surface is something that has not yet been imitated in the artificial model systems used to investigate lubrication in articular joints and intended to mimic the behaviour of articular cartilage. Additionally, this process can only take place when the surfaces are not immersed in synovial fluid. The knee joint, for example contains only between 0.5 and 4.0 mL of fluid.[50] Applying pressure to the large amounts of hyaluronan found at the surfaces of cartilage may lead to only water being expelled from the gel-like layer. As this mechanism of reforming the concentrated hyaluronan layer will occur when the joints are no longer in motion, it is not possible to draw conclusions concerning the role of the gel-like layer in reducing friction and wear during motion.

The very much larger concentrations of hyaluronan at the surface than in synovial fluid would certainly have a profound impact on the theories of elastohydrodynamic, squeeze film and multi-mode lubrication. These theories have concentrated on the viscosity, shear thinning and elasticity of synovial fluid.[76,77,86] Although a gel-like layer formed by concentration of hyaluronan at the surface would still allow fluid lubrication, the physical properties of this fluid would be very different to those of synovial fluid. For example, it has been shown that the viscosity of hyaluronan solutions increases from approximately 0.01 Pa.s at a concentration of 2 mg/mL to over 10 Pa.s at 20 mg/mL.[76] The retention of proteins in the concentrated, gel-like layer further increases the viscosity.[83,87] Elastohydrodynamic and squeeze-film theories tend to assume that the chemical composition of the liquid remains constant at the surface. However, the requirement on the synovial fluid to provide nutrients to the articular cartilage excludes this possibility.[18,76,86]

1.6.3. *Phospholipids*

Another early and very popular theory on how the articular cartilage is lubricated is that phospholipids are adsorbed onto the surface. Various suggestions have been made as to how these arrange themselves at the articular surface, either alone or together with proteins to form a membrane on the surface.[26,88,89] These propositions are generally based on the assumption that the articular surface is hydrophobic, as mentioned previously, which it is not.[30,76] The gel-like layer adsorbed onto the articular surface does become hydrophobic after removal from the joint but the articular surface below this does not.

The phospholipids that supposedly align to form bi- or multilayers at the surface are often referred to as "surface-active phospholipids". Although no phospholipid has been identified that is unique to the cartilage surface, no explanation has been offered as to how these phospholipids distinguish themselves from normal, run-of-the-mill, cell-membrane-forming phospholipids.[90–92] To compound the bizarre nature of the phrase "surface-active phospholipid", it was proposed that the proteoglycan lubricin carried the phospholipids to the surface where they formed layers, suggesting that there is no surface-activeness involved in the surface activity of these molecules.[93] Although the importance of providing nutrition to the articular cartilage was one of the earliest aspects to interest scientists, this imperative function of synovial fluid appears not to have been considered in many tribological studies.[12] This requirement on the system basically excludes the possibility of the formation of an impermeable membrane at the surface.

Finally, the formation of a solid dense phospholipid layer does, however, have a certain appeal to those who would view the system as behaving similarly to made-made, oil-lubricated systems, since it could allow analogous shifts in lubricating regime with Sommerfeld number. In contrast to aqueous, brush-like systems, a dense phospholipid layer would only display weak interactions with water, the strength depending on whether the head or tail is adsorbed onto the surface, leading to expulsion of the aqueous phase upon loading. This is, however, contrary to the way in which cartilage behaves.

1.6.4. *Lubricin*

By far the most thoroughly investigated candidate for friction reduction and wear protection is lubricin.[56,70] The recommended name for the protein component of lubricin assigned by protein databases such as uniprot (www.uniprot.org) is proteoglycan 4 (PRG4), since it is expressed by the PRG4 gene (Fig. 1.10). PRG4 is also known as megakaryocyte-stimulating

SMB   Mucin-like   PEX

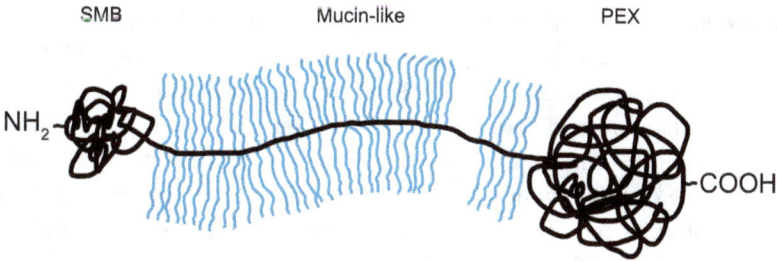

Fig. 1.10. Schematic representation of lubricin. The terminal globular proteins, somatomedin-B (SMB) and homeopexin (PEX) contain cysteine groups capable of forming sulphide bridges between lubricin units.[101]

factor (MSF), which is homologous to superficial zone protein (SZP). Lubricin, MSF and SZP differ in the glycosylation and sulphation of the glycosaminoglycan chains. Lubricin is referred to both as a proteoglycan and as a glycoprotein because although the protein backbone is attached to glycosaminoglycan chains, these are not sulphated, whereas in SZP, the glycosaminoglycan is sulphated.[56,70,94−96]

Lubricin is secreted by fibroblast-like synoviocytes, located in the inner layer (*intima*) of the *synovium*, and articular chondrocytes.[95] The observation that SZP is also expressed in chondrocytes in the transitional zone of cartilage as well as in synoviocytes has led to this being associated with lubrication.[97] Lubricin and SZP are expressed by the same gene and have similar protein structures but differ in post-translation modifications.[95] The low friction observed in the presence of lubricin has been attributed to the large amount of carbohydrate, which makes up approximately 50% of the weight of the glycoprotein.[74] The formation of a brush-like structure by lubricin adsorbed onto hydrophilic and hydrophobic surfaces has been proposed as the mechanism behind the friction-reducing properties.[64] This is consistent with the observation of very low friction coefficients in highly hydrophilic, charged and uncharged brush-like systems adsorbed onto both hard and soft surfaces.[79,98−100]

Lubricin has been found to attach to the transitional zone of cartilage.[102] Although the specific interactions could not be identified, it was shown that lubricin interacts with the transitional zone through the C-terminal domain (LUB-C), which reattached to the cartilage after removal of lubricin with 1.5 M NaCl aqueous solution. No interaction between the N-terminal domain and cartilage was detected. Additionally, binding to cartilage was significantly reduced by disruption of the disulfide bonds demonstrating that the secondary structure of the protein plays an important role in the interaction between lubricin and cartilage.[102] The N-terminal domain was shown to facilitate dimerization and thus it can be envisaged that lubricin forms a network by binding both to the cartilage and to itself.[102] The ability of lubricin to aggregate is also reduced by breaking the disulfide bonds.[63] Such a network of glycoprotein would allow the biomolecules required as nutrition to move between synovial fluid and the bulk of articular cartilage, perhaps by filtration of the synovial fluid

A number of strategies has been used to demonstrate that lubricin plays an important role in the tribology of articular cartilage, both directly in the articular joint by using a pendulum setup or by sliding articular cartilage against cartilage and glass.[51,103−107] Two pendulum systems, one designed to measure friction coefficients and the other to determine the effect of cyclic loading were used to investigate the role of lubricin in the tribology of knee joints. The knee joints were taken from mice that had been bred with and without the ability to produce lubricin.[103] These studies demonstrated the importance of lubricin not only in maintaining low friction coefficients but also in protecting the cartilage against wear under cyclic loading. A pendulum system was also used to show that the friction of healthy knee joints that were anterior cruciate ligament-deficient and thus contained significantly lower amounts of lubricin were higher than those of the healthy joints. Joints with mild osteoarthritis did not give significantly different friction coefficients when compared to the healthy joints and also contained similar amounts of lubricin.[105]

Tribological tests on natural articular cartilage sliding against cartilage have been used to investigate the possibility of cumulative effects from mixtures of hyaluronan, PRG4 (either lubricin or superficial zone protein) and phospholipid.[51] The results showed that both PRG4 and hyaluronan were able to reduce friction when compared to phosphate buffered saline (PBS). A combination of hyaluronan and PRG4 reduced the friction coefficient further. Addition of phospholipid did not result in a significant drop in friction, either alone or in combination with hyaluronan or PRG4.[51]

Studies on the exchange of PRG4 between cartilage and synovial fluid provide strong evidence that the PRG4 is tightly bound at the surface.[108] Disks of cartilage that included the superficial zone and the articular cartilage surface as well as disks from the middle zone were treated to determine the amount of PRG4. The articular surface stained more strongly for PRG4 than the surface of the middle zone. Treatment with sodium dodecyl sulphate, hyaluronidase, or reduction followed by alkylation all resulted in a significant depletion of PRG4 from the articular surface. Replenishment of the PRG4 occurred when the disks were subsequently stored in synovial fluid. The PRG4 content was highest in the disks that had been treated with hyaluronidase before replenishment in synovial fluid. Possibly, hyaluronan hinders the interaction of PRG4 with the cartilage. It is often claimed that PRG4 is bound to the articular surface, however, these results along with others suggest that PRG4 is the major component of the articular surface.[51,102,104,108] The investigations of Nugent-Derfus *et al.* found that PBS containing 2M NaCl had no significant effect on PRG4 at the surface, whereas experiments carried out by Jones *et al.* showed that lubricin was extracted by 1.5 M NaCl in PBS.[102,108] PBS alone has no effect on the articular surface even after soaking for a number of days.

Lubricin-mutant mice were bred to investigate the role of lubricin in protecting natural joints from wear.[28] With age, the joints of mice that lacked lubricin showed abnormal deposits of protein on the cartilage. It was concluded that lubricin not only preserves the integrity of the articular surface but also acts to maintain surface stiffness and the glycosaminoglycan content.[28] Additionally, the results suggested that lubricin is required to prevent the adsorption of the plasma proteins onto cartilage. As well as preventing non-specific protein adsorption, lubricin has also been shown to reduce cartilage-cartilage integration.[109,110]

1.7. Conclusions

While speculating on the role of the thick hyaluronan-protein layer at the surface, Balazs pointed out in 1974 that "the missing link in this argument, of course, is the complete lack of knowledge of the chemical composition of the cartilage surface layer in pathological conditions".[82] This situation remains unchanged. Numerous regions have been described as surfaces or have been assigned a name that suggests a surface. The tribological properties of these surfaces have all been investigated using various solutions as lubricant, a wide range of tribological conditions, and counter-surfaces of different chemistries. The resulting friction coefficients vary from approximately 0.002 up to 0.5. However, the interpretation of detected differences or similarities is complicated by the nature of the material and the absence of chemical specificity in friction coefficients. The material, articular cartilage, will always play a role in determining the friction coefficient, and therefore, in order to compare the friction in different experiments, not only the changes in the lubricant but also changes in the bulk cartilage or articular surface must be considered. Additionally, a change in the friction coefficient indicates some change in the tribological system, which may be lubricant, substrate, environment and/or tribological conditions: However, the same friction coefficient does not indicate that the tribological system remained unchanged. The observation that synovial fluid gives rise to lower coefficients of friction than PBS for cartilage vs cartilage in a particular measurement does not indicate that the mechanism of lubrication is the same as in the natural joint, but simply that synovial fluid is a better lubricant for this system than PBS.

The articular surface will, on the one hand, have special properties that enhance the lubrication of cartilage, such as participating in the mechanism of friction reduction and protecting the cartilage against wear. On the other hand, there are a number of other requirements on the articular surface in order to be fit for function, such as the ability to allow nutrients to permeate, resistance to non-specific protein adsorption and resistance to integration with the opposing cartilage surface. The large volume of literature on lubricin has shown that it is capable of fulfilling all of these requirements. Additionally, a mixture of lubricin and superficial zone protein would be consistent with all of the experimental chemical observation of the articular surface. It was found by Coles *et al.* that "lubricin appears to have a limited role in reducing the coefficient of friction on the cartilage surface of mammalian joints when only present on the articular surface. Despite this, it is apparently necessary for preservation of surface integrity, superficial stiffness, and GAG content.[28]

If lubricin does form the articular surface by forming intermolecular sulphide bonds then it can be envisaged that filtration of synovial fluid through this layer will lead to the observed enrichment of hyaluronan on the surface. The highly charged hyaluronan may then help to retain water at the surface during loading.

If lubricin is the main component of the articular surface, it could be argued that OA is not a problem of lubrication but a problem of damage to the material that is being lubricated. Many investigations into the tribological properties of natural cartilage focus on the lubricating abilities of synovial fluid and the friction between the contacting surfaces, without fully understanding the chemical nature of the surface. A more fruitful research direction might be to concentrate on finding a possibility to repair the articular surface rather than the current clinical practice of trying to supplement the already excellent lubrication that nature has provided.

References

1. G. A. Ateshian, C. T. Hung, The natural synovial joint: properties of carti-lage. *Proceedings of the Institution of Mechanical Engineers Part J-Journal of Engineering Tribology* 220(J8), 657–670 (2006).
2. A. Neville, A. Morina, T. Liskiewicz, Y. Yan, Synovial joint lubrication — does nature teach more effective engineering lubrication strategies? *P I Mech Eng C-J Mec* 221(10), 1223–1230 (2007).
3. J. Lane, C. Weiss, Review of articular-cartilage collagen research. *Arthritis Rheum* 18(6), 553–562 (1975).
4. J. Jurvelin, J. Arokoski, E. Hunziker, H. Helminen, Topographical variation of the elastic properties of articular cartilage in the canine knee. *J Biomech* 33(6), 669–675 (2000).
5. P. Roughley, Articular cartilage and changes in arthritis — Noncollagenous proteins and proteoglycans in the extracellular matrix of cartilage. *Arthritis Res* 3(6), 342–347 (2001).
6. F. Cajori, R. Pemberton, The chemical composition of synovial fluid in cases of joint effusion. *J Biol Chem* 76(2), 471–480 (1928).
7. V. C. Hascall, T. C. Laurent, Hyaluronan: Structure and Physical Proper-ties. http://www.glycoforum.gr.jp/science/hyaluronan/HA01/HA01E.html
8. C. Mörner, Chemische Studien über den Trachealknorpel. *Skand Arch Physiol* 1, 210 (1889).
9. W. Winter, Articles on the quantative composition of cartilage tissue. In *Biochemische Zeitschrift* Vol. 246, pp 10–28 1932.
10. K. Harpuder, Physical-chemical testings on the normal cartilage. In *Bio-chemische Zeitschrift*, Vol. 169, pp 308–319 1926.
11. W. Hunter, Of the structure and diseases of articulating cartilages. *Phil Trans* 42, 514 (1742).
12. L. Ito, The nutrition of articular cartilage and its method of repair. *Br J Surg* 12(45), 31–42 (1924).
13. M. A. MacConaill, Lubrication of mammalian joints. *Nature* 185(4717), 920–920 (1960).
14. P. Lewis, C. McCutchen, Lubrication of mammalian joints. In *Nature* Vol. 185, pp 920–921 1960.
15. L. Bian, M. Kaplun, D. Williams, D. Xu, G. Ateshian, C. Hung, Influence of chondroitin sulfate on the biochemical, mechanical and frictional properties of cartilage explants in long-term culture. *J Biomech* 42(3), 286–290 (2009).
16. V. Wright, D. Dowson, Lubrication and cartilage. In *Journal of Anatomy* Vol. 121, pp 107–118 1976.
17. B. Roberts, A. Unsworth, N. Mian, Modes of lubrication in human hip joints. *Annals of the Rheumatic Diseases* 41(3), 217–224 (1982).
18. T. Murakami, The lubrication in natural synovial joints and joint prostheses. *JSME Int J Iii-Vib C* 33(4), 465–474 (1990).
19. J. Katta, Z. Jin, E. Ingham, J. Fisher, Biotribology of articular cartilage-A review of the recent advances. *Med Eng Phys* 30(10), 1349–1363 (2008).
20. R. Crockett, Boundary lubrication in natural articular joints. *Trib Lett* 35(2), 77–84 (2009).

21. J. Clark, Variation of collagen fiber alignment in a joint surface-a scanning electron-microscope study of the tibial plateau in dog, rabbit, and man. in *J Orthopaed Res* Vol. 9, pp 246–257 (1991).

22. A. Jeffery, G. Blunn, C. Archer, G. Bentley, 3-Dimensional collagen architecture in bovine articular-cartilage. *J Bone Joint Surg Br* 73(5), 795–801 (1991).

23. C. Weiss, F. Shapiro, C. Trahan, K. Altmann, Tangential zone of articular-cartilage. In *Journal of Bone and Joint Surgery-American Volume* Vol. A 57, pp 584–584, (1975).

24. N. Kawai, E. Tanaka, T. Takata, M. Miyauchi, M. Tanaka, M. Todoh, T. van Eijden, K. Tanne, Influence of additive hyaluronic acid on the lubricating ability in the temporomandibular joint. *J Biomed Mater Res A* 70A(1), 149–153 (2004).

25. V. Mow, W. Lai, J. Eisenfeil, I. Riedler, Some surface characteristics of articular-cartilage 2. Stability of articular surface and a possible biomechanical factor in etiology of chondrodegeneration. *Journal of Biomechanics* 7(5), 457 (1974).

26. H. Higaki, T. Murakami, Y. Nakanishi, H. Miura, T. Mawatari, Y. Iwamoto, The lubricating ability of biomembrane models with dipalmitoyl phosphatidylcholine and gamma-globulin. *P I Mech Eng H* 212(H5), 337–346 (1998).

27. N. Wilson, D. Gardner, Influence of aqueous fixation on articular surface-morphology-a reflected light interference microscope study. *J Pathol* 131(4), 333–338 (1980).

28. J. Coles, L. Zhang, J. Blum, M. Warman, G. Jay, F. Guilak, S. Zauscher, Loss of cartilage structure, stiffness, and frictional properties in mice lacking PRG4. *Arthritis Rheum* 62(6), 1666–1674 (2010).

29. P. Kumar, M. Oka, J. Toguchida, M. Kobayashi, E. Uchida, T. Nakamura, K. Tanaka, Role of uppermost superficial surface layer of articular cartilage in the lubrication mechanism of joints. *J Anat* 199, 241–250 (2001).

30. R. Crockett, S. Roos, P. Rossbach, C. Dora, W. Born, H. Troxler, Imaging of the surface of human and bovine articular cartilage with ESEM and AFM. *Tribol Lett* 19(4), 311–317 (2005).

31. S. Graindorge, W. Ferrandez, E. Ingham, Z. Jin, P. Twigg, J. Fisher, The role of the surface amorphous layer of articular cartilage in joint lubrication. *P I Mech Eng H* 220(H5), 597–607 (2006).

32. M. Naka, K. Hattori, T. Ohashi, K. Ikeuchi, Evaluation of the effect of collagen network degradation on the frictional characteristics of articular cartilage using a simultaneous analysis of the contact condition. *Clin Biomech* 20(10), 1111–1118 (2005).

33. M. Naka, Y. Morita, K. Ikeuchi, Influence of proteoglycan contents and of tissue hydration on the frictional characteristics of articular cartilage. *P I Mech Eng H* 219(H3), 175–182 (2005).

34. K. Elsaid, G. Jay, M. Warman, D. Rhee, C. Chichester, Association of articular cartilage degradation and loss of boundary-lubricating ability of synovial fluid following injury and inflammatory arthritis. *Arthritis Rheum* 52(6), 1746–1755 (2005).

35. J. Jurvelin, D. Muller, M. Wong, D. Studer, A. Engel, E. Hunziker, Surface and subsurface morphology of bovine humeral articular cartilage as assessed by atomic force and transmission electron microscopy. *J Struct Biol* 117(1), 45–54 (1996).

36. R. Crockett, A. Grubelnik, S. Roos, C. Dora, W. Born, H. Troxler, Biochemical composition of the superficial layer of articular cartilage. *J Biomed Mater Res A* 82A(4), 958–964 (2007).

37. G. Meachim, D. Ghadiall.Fn; Collins, Regressive changes in superficial layer of human articular cartilage. *Ann Rheum Dis* 24(1), 23 (1965).

38. D. Gardner, D. McGillivray, Surface structure of articular cartilage-historical review. *Ann Rheum Dis* 30(1), 10 (1971).

39. C. McCutchen, Some surface characteristics of articular-cartilage. 1. Scanning electron-microscopy study and a theoretical model for dynamic interaction of synovial-fluid and articular-cartilage *J Biomech* 8(3–4), 261–261 (1975).

40. H. Forster, J. Fisher, The influence of continuous sliding and subsequent surface wear on the friction of articular cartilage. *P I Mech Eng H* 213(H4), 329–345 (1999).

41. Y. Sawae, T. Murakami, K. Matsumoto, M. Horimoto, Study on morphology and lubrication of articular cartilage surface with atomic force microscopy. *J Jpn Soc Tribologists* 45(2), 150–157 (2000).

42. C. Orford, D. Gardner, Ultrastructural histochemistry of the surface lamina of normal articular-cartilage. *Histochem J* 17(2), 223–233 (1985).

43. Z. Laver-Rudich, M. Silbermann, Cartilage surface-charge — A possible determinant in aging and osteoarthritic processes. *Arthritis Rheum* 28(6), 660–670 (1985).

44. E. Balazs, G. Bloom, D. Swann, Fine structure and glycosaminoglycan content of surface layer of articular cartilage. In *Federation Proceedings*, 1966; Vol. 25, (1813).

45. K. Mabuchi, Y. Tsukamoto, T. Obara, T. Yamaguchi, The effect of additive hyaluronic acid on animal joints with experimentally reduced lubricating ability. *Journal of Biomedical Materials Research* 28(8), 865–870 (1994).

46. T. Obara, K. Mabuchi, T. Iso, T. Yamaguchi, Increased friction of animal joints by experimental degeneration and recovery by addition of hyaluronic acid. *Clin Biomech* 12(4), 246–252 (1997).

47. F. Linn, E. Radin, Lubrication of animal joints. 3. Effect of certain chemical alterations of cartilage and lubricant. *Arthritis Rheum* 11(5), 674 (1968).

48. T. A. Schmidt, R. L. Sah, Effect of synovial fluid on boundary lubrication of articular cartilage. *Osteoarthr Cartilage* 15(1), 35–47 (2007).

49. J. J. Kwiecinski, S. G. Dorosz, T. E. Ludwig, S. Abubacker, M. K. Cowman, T. A. Schmidt, The effect of molecular weight on hyaluronan's cartilage boundary lubricating ability — alone and in combination with proteoglycan 4. *Osteoarthritis and Cartilage* 19(11), 1356–1362 (2011).

50. V. B. Kraus, T. V. Stabler, S. Y. Kong, G. Varju, G. McDaniel, Measurement of synovial fluid volume using urea. *Osteoarthritis And Cartilage* 15(10), 1217–1220 (2007).

51. T. A. Schmidt, N. S. Gastelum, Q. T. Nguyen, B. L. Schumacher, R. L. Sah, Boundary lubrication of articular cartilage — Role of synovial fluid constituents. *Arthritis Rheum* 56(3), 882–891 (2007).

52. J. P. Gleghorn, L. J. Bonassar, Lubrication mode analysis of articular cartilage using stribeck surfaces. *J Biomech* 41(9), 1910–1918 (2008).

53. R. Krishnan, M. Caligaris, R. Mauck, C. Hung, K. Costa, G. Ateshian, Removal of the superficial zone of bovine articular cartilage does not increase its frictional coefficient. *Osteoarthr Cartilage* 12(12), 947–955 (2004).

54. M. H. Naka, K. Hattori, K. Ikeuchi, Evaluation of the superficial characteristics of articular cartilage using evanescent waves in the friction tests with intermittent sliding and loading. *J Biomech* 39(12), 2164–2170 (2006).

55. H. E. Ozturk, K. K. Stoffel, C. F. Jones, G. W. Stachowiak, The effect of surface-active phospholipids on the lubrication of osteoarthritic sheep knee joints: friction. *Tribology Letters* 16(4), 283–289 (2004).

56. D. A. Swann, F. H. Silver, H. S. Slayter, W. Stafford, E. Shore, The molecular-structure and lubricating activity of lubricin isolated from bovine and human synovial-fluids. *Biochemical Journal* 225(1), 195–201 (1985).

57. E. D. Bonnevie, V. J. Baro, L. Wang, D. L. Burris, In situ studies of cartilage microtribology: roles of speed and contact area. *Tribology Letters* 41(1), 83–95 (2011).

58. J. Seror, Y. Merkher, N. Kampf, L. Collinson, A. J. Day, A. Maroudas, J. Klein, Articular cartilage proteoglycans as boundary lubricants: structure and frictional interaction of surface-attached hyaluronan and hyaluronan-aggrecan complexes. *Biomacromolecules* 12(10), 3432–3443 (2011).

59. J. Yu, X. Banquy, G. W. Greene, D. D. Lowrey, J. N. Israelachvili, The boundary lubrication of chemically grafted and cross-linked hyaluronic acid in phosphate buffered saline and lipid solutions measured by the surface forces apparatus. *Langmuir* 28(4), 2244–2250 (2012).

60. R. Tadmor, N. H. Chen, J. Israelachvili, Normal and shear forces between mica and model membrane surfaces with adsorbed hyaluronan. *Macromolecules* 36(25), 9519–9526 (2003).

61. R. Tadmor, N. H. Chen, J. N. Israelachvili, Thin film rheology and lubricity of hyaluronan solutions. *Biophysical Journal* 82(1), 799 (2002).

62. R. Tadmor, N. H. Chen, J. N. Israelachvili, Thin film rheology and lubricity of hyaluronic acid solutions at a normal physiological concentration. *Journal of Biomedical Materials Research* 61(4), 514–523 (2002).

63. B. Zappone, G. W. Greene, E. Oroudjev, G. D. Jay, J. N. Israelachvili, Molecular aspects of boundary lubrication by human lubricin: effect of disulfide bonds and enzymatic digestion. *Langmuir* 24(4), 1495–1508 (2008).

64. B. Zappone, M. Ruths, G. W. Greene, G. D. Jay, J. N. Israelachvili, Adsorption, lubrication, and wear of lubricin on model surfaces: polymer brush-like behavior of a glycoprotein. *Biophysical Journal* 92(5), 1693–1708 (2007).

65. J. Klein, Molecular mechanisms of synovial joint lubrication. *Proceedings of the Institution of Mechanical Engineers Part J-Journal of Engineering Tribology* 220(J8), 691–710 (2006).

66. M. Benz, N. H. Chen, J. Israelachvili, Lubrication and wear properties of grafted polyelectrolytes, hyaluronan and hylan, measured in the surface forces apparatus. *Journal of Biomedical Materials Research Part A* 71A(1), 6–15 (2004).

67. H. Watanabe, S. C. Cheung, N. Itano, K. Kimata, Y. Yamada, Identification of hyaluronan-binding domains of aggrecan. *Journal of Biological Chemistry* 272(44), 28057–28065 (1997).

68. T. E. Hardingham, Cartilage: Aggrecan — Link Protein — Hyaluronan Aggregates. http://glycoforum.gr.jp/science/hyaluronan/HA05/HA05E. html‡II

69. D. Chang, N. Abu-Lail, J. Coles, F. Guilak, G. Jay, S. Zauscher, Friction force microscopy of lubricin and hyaluronic acid between hydrophobic and hydrophilic surfaces. in *Soft Matter*, Vol. 5, pp 3438–3445 (2009).

70. G. Jay, Characterization of a bovine synovial-fluid lubricating factor. 1. Chemical, surface-activity and lubricating properties. *Connect Tissue Res* 28(1–2), 71–88 (1992).

71. M. Roba, M. Naka, E. Gautier, N. D. Spencer, R. Crockett, The adsorption and lubrication behavior of synovial fluid proteins and glycoproteins on the bearing-surface materials of hip replacements. *Biomaterials* 30(11), 2072–2078 (2009).

72. M. Roba, C. Bruhin, U. Ebneter, R. Ehrbar, R. Crockett, N. D. Spencer, Latex on Glass: an appropriate model for cartilage-lubrication studies? *Tribology Letters* 38(3), 267–273 (2010).

73. G. Jay, C. Cha, The effect of phospholipase digestion upon the boundary lubricating ability of synovial fluid. *J Rheumatol* 26(11), 2454–2457 (1999).

74. G. Jay, D. Harris, C. Cha, Boundary lubrication by lubricin is mediated by O-linked beta(1-3)Gal-GalNAc oligosaccharides. *Glycoconjugate J* 18(10), 807–815 (2001).

75. G. Jay, B. Lane, L. Sokoloff, Characterization of a bovine synovial fluid lubricating factor. 3. The interaction with hyaluronic acid. *Connective Tissue Research* 28(4), 245–255 (1992).

76. T. Murakami, H. Higaki, Y. Sawae, N. Ohtsuki, S. Moriyama, Y. Nakanishi, Adaptive multimode lubrication in natural synovial joints and artificial joints. *P I Mech Eng H* 212(H1), 23–35 (1998).

77. J. Hou, V. Mow, W. Lai, M. Holmes, An analysis of the squeeze-film lubrication mechanism for articular-cartilage. *J Biomech* 25(3), 247–259 (1992).

78. G. Ateshian, The role of interstitial fluid pressurization in articular cartilage lubrication. *J Biomech* 42(9), 1163–1176 (2009).

79. U. Raviv, S. Giasson, N. Kampf, J. F. Gohy, R. Jerome, J. Klein, Lubrication by charged polymers. *Nature* 425(6954), 163–165 (2003).

80. A. Ogston, J. Stanier, The physiological function of hyaluronic acid in synovial fluid — viscous, elastic and lubricant properties. *Journal of Physiology-London* 119(2–3), 244–252 (1953).

81. C. T. Stafford, W. Niiedermeier, H. L. Holley, W. Pigman, Studies on the concentration and intristic viscosity of hyaluronic acid in synovial fluids of patients with rheumatic diseases. *Annals of The Rheumatic Diseases* 23, 152–157 (1964).

82. E. A. Balazs, The physical properties of synovial fluid and the special role of hyaluronic acid. *Disorders of the Knee, JB LippincoU Co., Philadelphia*, 63–75 (1974).

83. J. R. Fraser, W. K. Foo, J. S. Maritz, Viscous interactions of hyaluronic acid with some proteins and neutral saccharides. *Annals of The Rheumatic Diseases* 31(6), 513–520 (1972).

84. P. Walker, A. Unsworth, D. Dowson, J. SIkorski, V. Wright, Mode of aggregation of hyaluronic acid protein complex on surface of articular cartilage. *Annals of the Rheumatic Diseases* 29(6), 591 (1970).

85. A. Maroudas, Hyaluronic acid films. *Proc Instn Mech Engrs* 181(3J), 122–124 (1966).

86. T. Pratt, D. James, The viscous squeeze-film behavior of shear-thinning fluids. *Journal of Non-Newtonian Fluid Mechanics* 27(1), 27–46 (1988).

87. J. Fraser, W. Murdoch, C. Curtain, B. WATT, Proteins retained with hyaluronic acid during ultrafiltration of synovial fluid. *Connective Tissue Research* 5(2), 61–65 (1977).

88. A.-M. Trunfio-Sfarghiu, Y. Berthier, M.-H. Meurisse, J.-P. Rieu, Role of nanomechanical properties in the tribological performance of phospholipid biomimetic surfaces. *Langmuir* 24(16), 8765–8771 (2008).

89. B. Hills, R. Crawford, Normal and prosthetic synovial joints are lubricated by surface-active phospholipid — A hypothesis. *J Arthroplasty* 18(4), 499–505 (2003).

90. A. Sarma, G. Powell, M. LaBerge, Phospholipid composition of articular cartilage boundary lubricant. *Journal of Orthopaedic Research* 19(4), 671–676 (2001).

91. J. Lombard, P. Lopez-Garcia, D. Moreira, The early evolution of lipid membranes and the three domains of life. *Nat Rev Microbiol* 10(7), 507–515 (2012).

92. I. M. Cristea, M. Degli Esposti, Membrane lipids and cell death: an overview. *Chem Phys Lipids* 129(2), 133–160 (2004).

93. B. Hills, Boundary lubrication *in vivo*. *P I Mech Eng H* 214(H1), 83–94 (2000).

94. G. Jay, U. Tantravahi, D. Britt, H. Barrach, C. Cha, Homology of lubricin and superficial zone protein (SZP): products of megakaryocyte stimulating factor (MSF) gene expression by human synovial fibroblasts and articular chondrocytes localized to chromosome 1q25. *Journal of Orthopaedic Research* 19(4), 677–687 (2001).

95. G. D. Jay, D. E. Britt, C. J. Cha, Lubricin is a product of megakaryocyte stimulating factor gene expression by human synovial fibroblasts. *Journal of Rheumatology* 27(3), 594–600 (2000).

96. D. Swann, R. Hendren, E. Radin, S. Sotman, E. Duda, The lubricating activity of synovial fluid glycoproteins. *Arthritis and Rheumatism* 24(1), 22–30 (1981).

97. T. Niikura, A. H. Reddi, Differential regulation of lubricin/superficial zone protein by transforming growth factor beta/bone morphogenetic protein superfamily members in articular chondrocytes and synoviocytes. *Arthritis Rheum* 56(7), 2312–2321 (2007).

98. S. Lee, M. Muller, M. Ratoi-Salagean, J. Voros, S. Pasche, S. M. De Paul, H. A. Spikes, M. Textor, N. D. Spencer, Boundary lubrication of oxide surfaces by Poly(L-lysine)-g-poly(ethylene glycol) (PLL-g-PEG) in aqueous media. *Tribology Letters* 15(3), 231–239 (2003).

99. S. Lee, N. D. Spencer, Materials science — sweet, hairy, soft, and slippery. *Science* 319(5863), 575–576 (2008).

100. M. Muller, S. Lee, H. A. Spikes, N. D. Spencer, The influence of molecular architecture on the macroscopic lubrication properties of the brush-like co-polyelectrolyte poly(L-lysine)-g-poly(ethylene glycol) (PLL-g-PEG) adsorbed on oxide surfaces. *Tribology Letters* 15(4), 395–405 (2003).

101. D. Rhee, J. Marcelino, M. Baker, Y. Gong, P. Smits, V. Lefebvre, G. Jay, M. Stewart, H. Wang, M. Warman, J. Carpten, The secreted glycoprotein lubricin protects cartilage surfaces and inhibits synovial cell overgrowth. *J Clin Invest* 115(3), 622–631 (2005).

102. A. R. C. Jones, J. P. Gleghorn, C. E. Hughes, L. J. Fitz, R. Zollner, S. D. Wainwright, B. Caterson, E. A. Morris, L. J. Bonassar, C. R. Flannery, Binding and localization of recombinant lubricin to articular cartilage surfaces. *J Orthopaed Res* 25(3), 283–292 (2007).

103. E. I. Drewniak, G. D. Jay, B. C. Fleming, L. Zhang, M. L. Warman, J. J. Crisco, Cyclic loading increases friction and changes cartilage surface integrity in lubricin-mutant mouse knees. *Arthritis and Rheumatism* 64(2), 465–473 (2012).

104. G. D. Jay, J. R. Torres, D. K. Rhee, H. J. Helminen, M. M. Hytinnen, C.-J. Cha, K. Elsaid, K.-S. Kim, Y. Cui, M. L. Warman, Association between friction and wear in diarthrodial joints lacking lubricin. *Arthritis Rheum* 56(11), 3662–3669 (2007).

105. E. Teeple, K. A. Elsaid, B. C. Fleming, G. D. Jay, K. Aslani, J. J. Crisco, A. P. Mechrefe, Coefficients of friction, lubricin, and cartilage damage in the anterior cruciate ligament-deficient guinea pig knee. *J Orthopaed Res* 26(2), 231–237 (2008).

106. J. Gleghorn, A. Jones, C. Flannery, L. Bonassar, Boundary mode lubrication of articular cartilage by recombinant human lubricin. in *J Orthopaed Res*, Vol. 27, pp 771–777 (2009).

107. J. Gleghorn, A. Jones, C. Flannery, L. Bonassar, Alteration of articular cartilage frictional properties by transforming growth factor beta, interleukin-1 beta, and M. oncostatin *Arthritis Rheum* 60(2), 440–449 (2009).

108. G. E. Nugent-Derfus, A. H. Chan, B. L. Schumacher, R. L. Sah, PRG4 exchange between the articular cartilage surface and synovial fluid. *J Orthopaed Res* 25(10), 1269–1276 (2007).
109. C. Englert, K. McGowan, T. Klein, A. Giurea, B. Schumacher, R. Sah, Inhibition of integrative cartilage repair by proteoglycan 4 in synovial fluid. *Arthritis Rheum* 52(4), 1091–1099 (2005).
110. D. Schaefer, D. Wendt, M. Moretti, M. Jakob, G. Jay, M. Heberer, I. Martin', Lubricin reduces cartilage-cartilage integration. *Biorheology* 41(3–4), 503–508 (2004).

Chapter 2

Sticky and Slippery: Interfacial Forces of Mucin and Mucus Gels

Seunghwan Lee

Department of Mechanical Engineering
Technical University of Denmark
seele@mek.dtu.dk

2.1. Introduction

Mucosa lines internal organ tissues and body cavities and plays a protective role by secreting slimy and adherent mucus layers. Mucus layers are hydrogel networks formed from macromolecules called mucins that are secreted by glands, together with other biomolecules, such as immunoglobulin, lipids, salts, and water.[1-3] As shown in the Fig. 2.1, mucin is large-molecular-weight glycoprotein, constructed from a linear protein backbone with heavy glycosylation of the central part in a tandem pattern. Carbohydrate chains account for 70–80% of the molecular weight of mucin and impart the hygroscopic nature to mucus gels. Mucins may exist as a single unit, i.e. in the form of single chain from C to N termini, but can form aggregates via disulfide bonding or via hydrophobic interactions.

In the tribology community, mucin is probably best known as a lubricant for biological systems. This is readily appreciated from our everyday experiences, such as the lubricating role of saliva for speech and mastication of food in mouth, the smooth passage of a food bolus through the esophagus and gastrointestinal tracts, and tears for smooth eye blinking. In fact, tribological studies have shown that isolated molecular mucins show clear lubricating effects at various sliding interfaces.[4-9] The idea of mucin/mucus being a lubricant additive partly derives from the fact that the structure of mucin is similar to those of other biomacromolecules that are involved in biolubrication, such as lubricin in synovial fluid[10-13] and aggrecan/cartilage

potential aggregation through
hydrophobic interaction

or   disulfide bonding

glycosylated
region

unglycosylated region

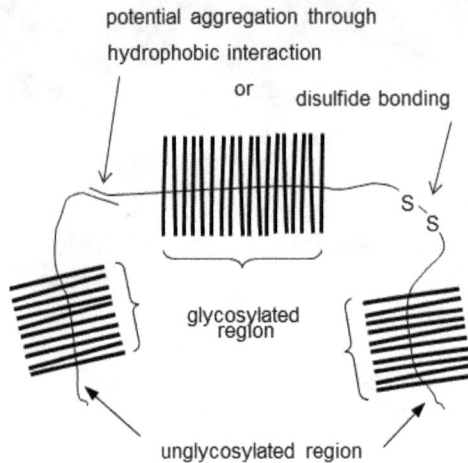

Fig. 2.1.   Schematic illustration of mucin and mucus aggregates.

tissues of articular joints.[14-17] Since water is the exclusive base fluid
for biological systems, it is a natural choice to decorate proteins with
hydrophilic moieties, such as sugars, to interact with and entrain water,
and to lubricate biological tissues. Another aspect of mucins is that their
biological functions go far beyond than lubrication, and include bioadhe-
sion, cell signal transduction,[18] diagnostic markers in cancer,[19,20] bacterial
inoculation,[21] immune response,[22] inflammation,[23,24] and tumorigenesis.[25]
Interestingly, these other functions of mucin/mucus are mostly related to
the *specific recognition* and *adhesive* properties of mucins with counter
surfaces. In other words, mucin and mucus gels display not only slippery,
lubricious properties, but also sticky, adhesive properties,[26,27] depending on
the nature of the interacting counter surfaces. Recent studies have begun
to characterize the adhesive forces arising from mucin layers, for instance,
mucoadhesion[28-33] and ligand-receptor interaction.[34-40]

Based upon the notion that mucin/mucus gels display "dual" properties
at the interface, in this chapter, it is attempted to review previous studies
that have focused on interfacial forces involving mucins and mucus gels,
their behavior being adhesive, repulsive, or even slippery. The approaches
and motivations of these studies are very diverse, and include mucoadhesion for drug delivery, understanding the lubricating properties of saliva
in food science and odontology, masticatory food consumption in food science, smooth gliding of endoscopy in biomedical science, and fundamental
understanding of mucin lubrication behavior in biotribology. Nevertheless,

common to all these studies is that interfacial forces are the principal parameter to be quantified and understood. For this reason, the instrumental approaches employed are those that the tribology community is familiar with, involving the surface forces apparatus (SFA), the atomic force microscope (AFM), and various tribometers. There is therefore a strong potential that the tribology community will be most interested in the topic, and be able to contribute to further studies.

This chapter is composed of two main parts, covering molecular mucins (Sec. 2.2) and mucus gels/mucosa (Sec. 2.3). Molecular mucins offer an opportunity to understand the fundamental force-interaction mechanisms at interfaces, while mucus gels offer an opportunity to understand interacting forces in systems of physiological relevance.

2.2. Molecular Mucins

2.2.1. *Repulsion*

To date, only a few types of mucins have been studied by SFA, namely, bovine submaxillary mucin (BSM),[41–44] porcine gastric mucin (PGM),[7,45] and rat gastric mucin (RGM).[45] As a substrate, many of these studies employed freshly cleaved mica,[41–44] while others employed hydrophobized mica.[7,45] Sub-topics concerning surface forces between mucin layers include the fundamental interactions between mucin-coated layers[7,41–45] and the influence of solution environmental parameters, such as pH and salts.[7,42,46] Additionally, specific questions concerning mucins include the influence of purification on surface-force profiles[44] and resistance to surfactants by the formation of protective layers.[43]

Interaction between mucin and mucin

Proust *et al.* were among the first to report measurements of the surface forces between BSM layers adsorbed on mica surfaces.[41,42] In a separate study,[48] the authors characterized the adsorption behavior of BSM onto various surfaces from aqueous solution, for example, silicone, mica, PVP-grafted mica, polyethylene films, and oxidized PE films. Due to its polyampholytic nature, BSM can readily adsorb onto even negatively charged mica surfaces despite its overall polyanionic character; N-termini and unglycosylated amino acid moieties carrying positive charges are considered to be responsible for the interaction with negatively charged mica surfaces. Adsorption onto hydrophilic mica surfaces by mucins[7,43,44] and by other polyanionic glycoproteins such as lubricin from synovial fluid[10-13] was later reported by many other groups. However, it should also be noted that the adsorption of mucins is not generally occurring to all hydrophilic/ negatively charged surfaces, since the amount of adsorption of BSM onto silica is significantly smaller compared to its adsorption onto mica.[44] This implies that adsorption of mucin onto mica surface is not entirely non-specific, and some other interaction mechanisms might play a role. In general, adsorption of proteins onto hydrophilic surfaces induces insignificant denaturation,[49] but weak anchoring stability is unavoidable due to the few interacting points. Further, it takes significant time to establish the equilibrium (in fact, the equilibrium was not reached until 24 hours for BSM on a mica surface[41,42,48]), and a significant portion of the adsorbed molecules could be removed by rinsing with aqueous buffer solution.

Variation of the BSM concentration, 0.1 mg/ml and 0.2 mg/ml, and salt concentration in buffer, 0.001 M NaCl and 0.15 M NaCl,[42] revealed that both parameters significantly influence the surface adsorption and surface forces of BSM; the interaction in low BSM concentration and low-salt buffer is characterized by long-range attractive forces caused by bridging, hysteresis between compression/decompression cycles, and exponential, repulsive steric interaction. However, long-distance attractive forces disappear upon employing 0.2 mg/ml of BSM and hysteresis completely disappears when 0.15 M NaCl solution is used in addition.

Later studies by Dedinaite et al.[43] and Lundin et al.[44] have confirmed a few common features for the surface forces between BSM layers on mica. Firstly, DLVO theory cannot fully explain the repulsive regime, and repulsion is attributed primarily to steric effects. Secondly, the first compression is generally more repulsive than subsequent ones, implying that structural rearrangement initially occurs; long loops and tails of mucin may extend out from the surface into the bulk solution, and these tails and loops become irreversibly compressed during the first cycle.

Harvey et al. have employed PGM ("Orthana", extensive analytical information on this PGM is available in another study[50]) for surface-force studies by SFA.[7] In this work, both hydrophilic and hydrophobized mica surfaces (prepared by modification with a surfactant layer, "STAI") have been employed. The concentration of PGM was 1 mg/ml. The onset of the repulsion at a hydrophilic mica surface, 40–70 nm, was at lower separation than for the case of BSM mentioned above (110 nm–170 nm[42]), presumably because of the lower molecular weight of Orthana PGM (ca. 0.55 MDa[50]). Similarly to BSM, a slight hysteresis between the first compression and decompression was observed, even at this concentration. However, no difference was observed between the first and subsequent compressions/decompression cycles. The adsorption behavior of PGM onto hydrophobized mica surfaces is markedly different; adsorption equilibrium is reached quickly, a higher amount being adsorbed ($\sim 2.0\,\text{mg/m}^2$ for hydrophobized mica and $\sim 1.3\,\text{mg/m}^2$ for hydrophilic mica, respectively[7]), hydrophilic moieties are exposed to the bulk solution, and finally little desorption is observed upon rinsing. Nevertheless, the integrity of PGM onto both hydrophilic and hydrophobized mica surfaces is quite evident, since shear stress up to $300\,\mu\text{N}$ (to be further discussed further below) did not alter the surface force profiles in either case. Furthermore, the surface-force profiles of PGM on hydrophobized mica were very similar to those on hydrophilic mica, except that compression/decompression hysteresis was completely absent for the case of hydrophobized mica.

Malmsten *et al.* have also employed hydrophobized mica (by a Langmuir-Blodgett film) to investigate surface-force effects of rat gastric mucin (RGM) and PGM.[45] Both gastric mucins have been isolated and purified from tissues by the authors. While PGM is larger than RGM in both molecular weight (15 MDa vs. 10 MDa, respectively) and radius of gyration (R_g) (210 nm vs. 190 nm, respectively), the amount adsorbed onto the surface is rather higher for RGM than PGM ($3.5 \pm 1.5\,\text{mg/m}^2$ and $3.0 \pm 1.5\,\text{mg/m}^2$, respectively, as determined by SFA), and most of all, the layer thickness measured by compression is distinctively smaller for PGM than RGM (3.5 nm vs. 10–13 nm). This is mainly attributed to a higher charge density and stronger polyelectrolyte characteristics for PGM than for RGM and a consequently flatter conformation on the surface. It is also notable that adsorption and surface-force behavior of PGM in this work differs in many ways from those of the aforementioned PGM study by Harvey *et al.*[7] A primary difference between these two studies is that PGM in this work is much larger than Orthana PGM, both in molecular weight (15 MDa[45] vs. 0.55 MDa[7]) and R_g (150 nm[45] vs. 55 nm[7]).

Influence of salts

Influence of salts on the surface forces of mucin-covered mica surfaces has briefly been commented upon in each case above. Perez *et al.*[42] showed higher adsorption of BSM from 0.15 M NaCl solution than from 1×10^{-3} M M NaCl solution and consequently stronger repulsive forces (Fig. 2.2). On the other hand, Harvey *et al.*[7] reported that high-salt buffer (0.1 M NaCl

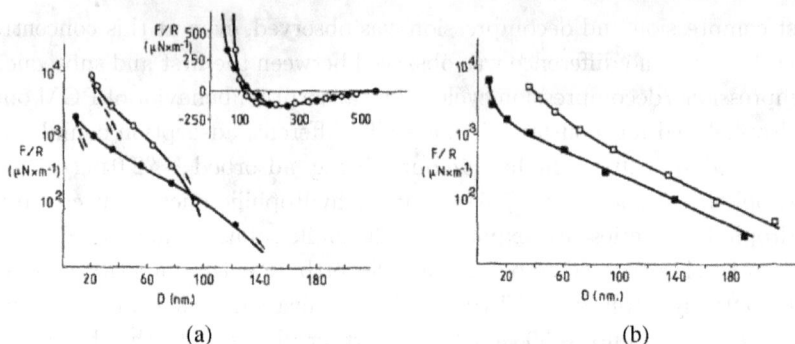

Fig. 2.2. Surface-force profiles between two mica surfaces covered with adsorbed BSM for 3 h (a) (•) BSM 0.1 g/liter + NaCl 10^{-3} M (case A), (○) 0.05 g/liter + NaCl 0.15 M, (b) (□) 0.2 g/liter + NaCl 10^{-3} M, (■) 0.05 g/liter + NaCl 0.15 M (From Ref. 42).

or NaNO$_3$) leads to a compaction of the PGM layer and reduced repulsive forces. This contrast can be attributed to the latter study using the high-salt buffer solution only for rinsing purpose (adsorption of mucin was carried out from salt-free, pure water), whereas the former study used the high-salt buffer solution throughout the measurements, starting from the adsorption of mucins. Another important aspect of the salt influence on the surface forces between mucin layers has been studied by Pettersson et al.[46] They characterized the forces acting between two 1-hexadecanethiol-coated gold surfaces (10 μm gold particle and gold QCM crystal), i.e. two hydrophobic surfaces, coated with BSM, across various salted aqueous buffer solutions by means of AFM. In comparing 1 mM CaCl$_2$ (Debye length = 5.6 nm) and 30 mM NaNO$_3$ solutions (Debye length = 1.8 nm), the layer extension is expected to be larger in 1 mM CaCl$_2$ than in 30 mM NaNO$_3$, if electrostatic screening effect is dominant. However, the opposite behavior was observed in surface-force profiles; the authors suggested that Ca^{2+} ions bind to the BSM layers, causing the layers to compact. In other words, specific interactions of Ca^{2+} ions with BSM overwhelm the total ionic strength in this case. Significant contraction of the tails, even in dilute salt solutions (1 mM), was commonly observed with multivalent cations, such as Ca^{2+} and La^{3+}.

Influence of purification

Another important issue for surface-force studies specific to mucin layers is the influence of "contaminants" and purification. Lundin et al. have employed commercial BSM (Sigma-Aldrich) and carried out extensive chromatographic purification to remove non-mucin components.[44] While commercially available mucins provided are having undergone certain purification processes, it is well known that these products still contain a significant amount of non-mucin species. Details on the purification by Lundin *et al.* are available in Ref. 51. Albumin was identified as the major "contaminant". For this reason, the authors have compared the compressed layer thicknesses of (a) purified, (b) "as-received", (c) BSA, (d) BSM + BSA (1:1), and (e) BSM + BSA (10:1), and they were measured to be (a) 10, (b) 30–35, (c) 50, (d) 120, and (e) 30–35 Å, respectively (adsorbed from 0.05 mg/ml glycoprotein solutions, except for 10:1 mixture, where adsorption occurred from the mixture of 0.05 mg/ml BSM + 0.005 mg/ml BSA). It is interesting to note that the purified BSM reveals the smallest adsorbed mass, most compact layer, and weakest repulsive forces. The long-range interaction starts from ca. 50 nm, which is much smaller than observed for BSM by Perez *et al.*[42] It is also notable that while the compressed film thickness of "as-received" BSM and the 1:0.1 purified BSM:BSA mixture are nearly identical, detailed surface force profiles of the two samples are still different, as shown in Fig. 2.3.

One major difference is that the hysteresis between compression and decompression cycle is observed for the "as-received" BSM, but not for the

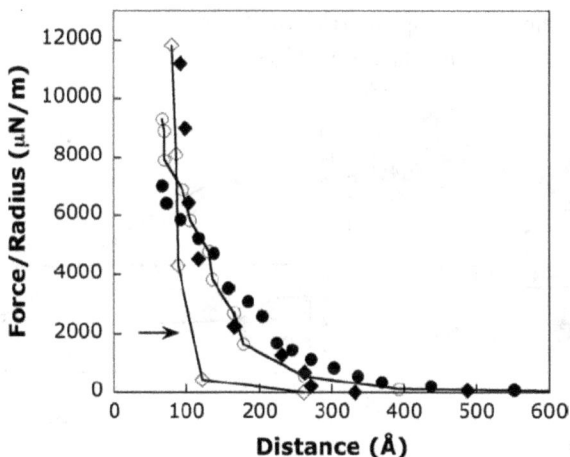

Fig. 2.3. Normalized force as a function of surface separation between two mica surfaces pre-coated with as received BSM (diamonds) and a BSM:BSA 1:0.1 mixture (circles) across a PBS20 buffer solution. Filled and unfilled symbols represent the force measured on approach and separation, respectively. The lines are drawn only for guidance (from Ref. 44).

BSM:BSA mixture, which may indicate that a simple mixture of BSM and BSA do not reproduce the behavior of an as-received BSM sample.

Surface forces between mucin + surfactants

Dedinaite *et al.* have not only studied BSM-BSM layer interactions, but also the behavior of a composite layer of BSM and chitosan and further interactions when this layer is exposed to surfactants (SDS).[43] Chitosan was selected as the partner to form the composite layer with BSM for its protective properties against SDS. Combined studies with ellipsometry and QCM-D revealed that adsorption of chitosan on top of a mucin layer results in a small increase in adsorbed mass, yet a decrease in film thickness. This indicates that compaction occurs upon adsorption of chitosan on top of BSM layer (see Fig. 2.4(a) for schematic illustration).

The composite layer of BSM + chitosan was resistant to rinsing with an inert salt solution or SDS, demonstrating that the chitosan layer is firmly attached to the mucin. Incorporation of chitosan into the adsorbed BSM layer and its protective role are observed from the surface forces as well. Upon formation of the BSM + chitosan composite layer, less repulsive forces were detected. Nevertheless, exposure of this layer to SDS solution (1 cmc) did not alter the surface-force profiles, unlike the case of the BSM layer alone. The protective property of chitosan for BSM against SDS derives from the intrinsic properties of the chitosan film itself; exposure of pure chitosan adsorption layers to SDS solutions of concentrations up to

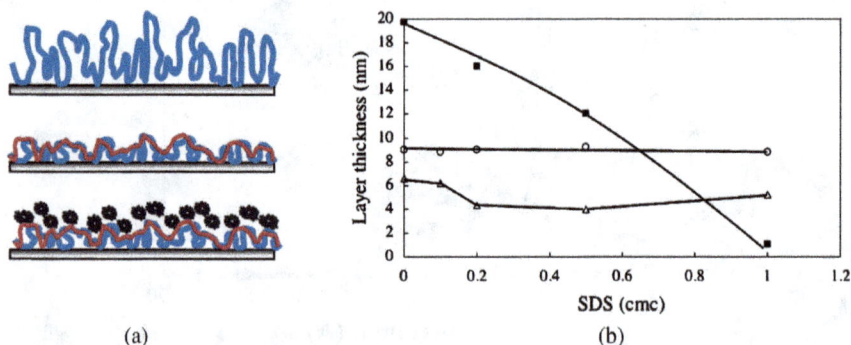

(a)   (b)

Fig. 2.4. (a) Cartoon showing, from top down, the mucin adsorption layer, mucin collapsed by chitosan, and the mucin-chitosan layer with adsorbed SDS (b) Polyelectrolyte layer thickness as a function of SDS concentration, measured by SFA between mica surfaces (symmetrical system) at $F/R = 10$ mN/m: mucin (■), mucin-chitosan (○), chitosan (△). The background electrolyte is 30 mM $NaNO_3$ (from Ref. 43).

1 cmc results in virtually no desorption, but merely a strong compaction (Fig. 2.4(b)).[43,47] Petterson *et al.* repeated the same experiments i.e. the interaction between the composite layers of BSM and chitosan in the presence and absence of SDS by AFM at the interface of AFM tip (silica) and substrate (mica).[8] Basically, compaction of the BSM layer upon formation of a composite layer with chitosan and the protective role of chitosan against SDS are reproduced. However, it is notable that, compared to SFA measurements, long-range adhesive forces were more readily detected from the composite layer by AFM. Furthermore, the protective role of chitosan against SDS was not observed after applying shear stress (to be discussed further below), which had not been explored in the previous SFA work.[43] This could be due to the different sensitivities of SFA and AFM, but also could result from the different coverage and roughness of a silica surface and an AFM tip compared to the mica surface.

2.2.2. Adhesion

Adhesion between mucin and defined solid substrates

Berry *et al.* have employed human ocular mucins, and studied the interaction with a mica surface in neutral aqueous buffer environment by AFM.[52] The authors modified AFM tips with mucins via the reaction between a succinimide-SAM on an AFM tip and the primary amines on the mucins. Diverse, multiple force pull-off events were observed. The authors grouped the results into two main types of pull-off events: short-length jumps in the force-distance curve denoted as "detachment events" (occurring at distances much shorter than the known lengths of mucin molecules[53]) and longer duration "stretching events". Adhesion energy increased in the presence of the divalent cations (e.g. $NiCl_2$) and with dwell time, but was not affected by the approach speed. The interaction between human ocular mucin and a positively charged surface (APTES-mica substrate) was also studied by the same group, especially during the first few seconds of interaction.[54] The motivation of this work was related to an old concept that negatively charged mucin molecules introduced to a positively charged surface are kinetically trapped and bind very rapidly. This study has revealed that, in fact, significantly different kind of adsorption kinetics of mucins occur during the first three seconds of exposure to the surface and proved that molecular rearrangement is required for several seconds before trapping occurs.

Mucoadhesion

Mucoadhesive polymers have the potential to be employed as drug carriers for localized delivery and/or to prolong the residence time of a delivery device. They rely upon adhesive interactions with mucus layers.[55] Conventional techniques to measure the mucoadhesive properties include tensile-force measurements,[56–58] fluorescence probe measurements,[59] flow-property measurements of polymer particles in a channel filled with mucous gel,[60,61] and rheological methods.[62] An outstanding advantage of SFA and AFM as assessment techniques for mucoadhesion is that they provide a *direct* measurement of mucoadhesive forces between polymer and mucin layers on molecular level. In addition, it is possible to clearly distinguish trends in mucoadhesive forces as a function of environmental variables, such as pH and ionic strength, as well as sampling speed. However, preparation for experiments is generally more complicated than in conventional methods and reproducibility is somewhat limited due to variations in thickness and topography of polymer and mucin layers.[31]

Efremova *et al.* were one of the first groups who employed SFA to investigate the interactions between grafted poly(ethylene glycol) (PEG) (Mw 2000) and BSM.[28] Both adsorbed and soluble BSM were employed in order to emulate the physiological condition of gastric organs. The adhesive interactions between PEG and BSM layers were identified, both in the presence and absence of excess BSM (0.2 mg/ml). Even in the case where both surfaces are terminated with PEG layers, which leads to pure repulsion across buffer solution, soluble mucins were observed to mediate bridging attractions between the PEG layers. However, the magnitude of this interaction is not strong; as SPR measurements of mucin adsorption onto a supported PEG monolayer showed, all BSMs are completely rinsed away with HEPES buffer solution. The attraction force is enhanced by nearly five times at pH 2 than pH 7.2, supporting the idea that hydrogen bonding is chiefly responsible for mucoadhesion between BSM and PEG layers. In understanding the mucoadhesive properties at low pH, it needs to be considered that PEG also goes through changes in water-solubility and conformation at low pH.[63,64] Weaker adhesive forces at neutral pH are a problem for the application of mucoadhesive polymers as drug carriers targeting at intestines. Zhu *et al.* have employed the copolymer of N-isopropylacrylamide and glycidylacrylamide (NIPAM-Gly) as mucoadhesives, to take advantage of the fact that this copolymer forms hydrogen bonds at pH 5.0 (near the pKa of carboxylate), i.e. close to neutral pH.[29] To focus on the interaction of NIPAM-Gly chains with free, soluble BSM, the surface force profiles were obtained between NIPAM-Gly brushes with or without BSM molecules across them. The presence of free BSM led to a significant increase in both the range and the magnitude of the repulsion, which indicates the adhesion between free BSM and NIPAM-Gly layers.

Contrary to expectations after these first papers,[28,29,55] the study of mucoadhesion by means of SFA did not continue to be very active thereafter. Instead, AFM has emerged as a tool to characterize mucoadhesive interactions between mucins and polymers. J. Cleary *et al.* have studied mucoadhesion properties of Pluronic-PAA copolymer against BSM by means of AFM.[30] A 20 μm glass microsphere modified with a layer of Pluronic-PAA copolymer was attached to the AFM cantilever to form a colloidal probe. A BSM layer was prepared by attaching a layer of mucin onto an epoxy adhesive layer deposited on a glass plate. The adhesive force is observed to be highest at pH values of 4–5, coinciding with the pKa of the carboxylic groups, and decreased at both lower and higher pH values (Fig. 2.5(a)).

A few factors must be considered to understand the pH-dependence of the adhesive forces. Since hydrogen bonding is significantly contributing to mucoadhesion, higher adhesive forces are expected when the existing acidic groups are mostly protonated. In this regard, the highest adhesion should be observed at pH 2, where all of the acidic groups, both from Pluronic-PAA copolymers (pKa ≈ 4.95) and BSM (pKa ≈ 2.6) are protonated. Upon increasing pH, electrostatic repulsion is certainly increasing due to increasing deprotonation of acidic groups. At the same time, however, a greater degree of entanglement and interpenetration of polymer chains with mucin molecules is also possible, due to the more open structure caused by internal electrostatic repulsion. Thus, increasing adhesive forces with increasing pH can be explained (within the range of pH 2 to 5). As pH is further raised, however, not only charge-charge repulsion becomes dominant, but also the polymer structure becomes too loose due to increased water absorption, such that adhesive forces decrease again. Again, this reduction in adhesive forces under neutral pH is a problem for drug-delivery vehicles for intestines. Iijima *et al.* have studied the interaction between a p(NIPAAm) (core) /p(MMA-g-EG) (shell) nanogel particle and a gastric mucin layer (species not described) dispersed on APTS-modified mica plates.[31] In contrast to the work by Cleary *et al.*,[28] adhesive forces were observed to be *minimum*

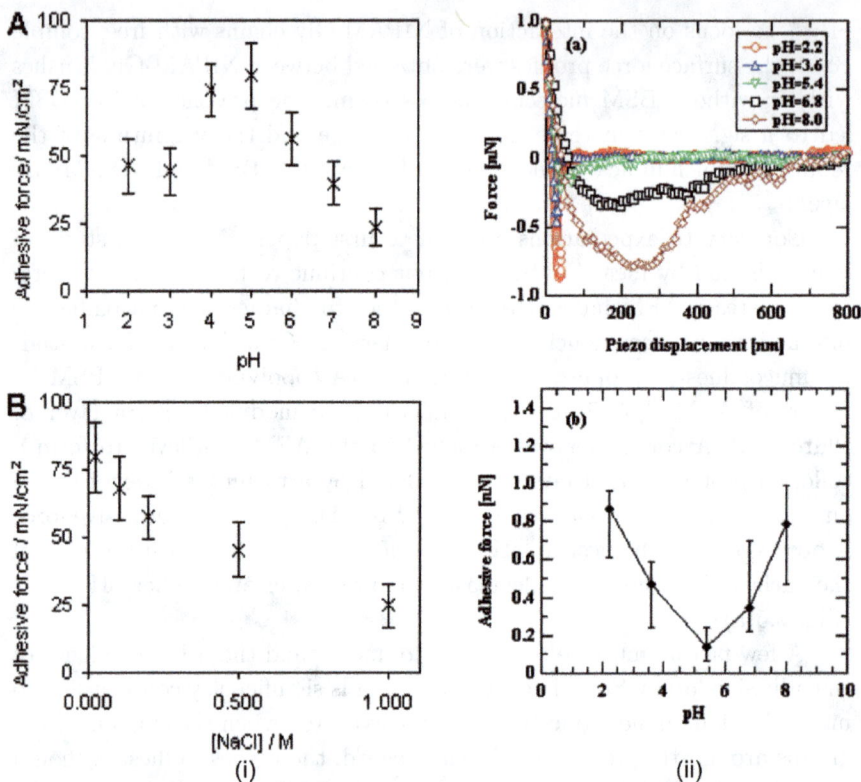

Fig. 2.5. (i) Adhesive force between mucin and Pluronic-PAA as a function of (A) pH and (B) NaCl concentration (from Ref. 30) (ii) (a) Effect of pH on the interaction between a nanogel particle and a mucin layer during the separation process and (b) the magnitude of the maximum adhesive force at various pH values (from Ref. 31).

when pH was 5.4 (IEP of nanogel particle is ca. pH 4), and increased under more neutral or acidic conditions (see Fig. 2.5(b)). The authors ascribed (i) minimum adhesive forces at pH 5.4 to charge-charge repulsion, (ii) increasing adhesive forces towards low pH values to the dominance of hydrogen bonding between protonated surfaces, and finally (iii) increasing adhesive forces at neutral pH to the swollen p(MAA-*g*-EG) shell network structure, similarly with the explanation for the maximum adhesive forces shown at pH 5 by Cleary *et al.*[30] In fact, force-distance curves at pH 2 and 8 are similar in pull-off forces, i.e. adhesive forces, but significantly different in "pull-off lengths". At pH 2, the adhesive interaction is short ranged, whereas at pH 8, the interaction is much longer ranged, supporting the argument that the difference is caused by the inclusion of entanglement (pH 8) or not (pH 2), respectively (see Fig. 2.5(b)). Since the IEPs of the nanogel particle (4.0) and the gastric mucin layer (2.6)[65] are similar to those of Pluronic-PAA (4.95) and BSM for the previous study (2.6),[30] the different pH dependence shown in these two studies is not consistent in view of electrostatic interaction. A notable difference between the two systems, however, is that pH-dependent moieties on nanogel particle reside at the outermost layer (PMAA-g-EG shell layer), whereas those of Pluronic-PAA copolymer is more shielded inside, such that the detailed mechanisms of swelling, and thus entanglement with mucin layers might be fairly different for these two polymers. Furthermore, the types of mucins employed for the two studies are different.

Carton *et al.* have shown the possibility of using a DOPA moiety as a powerful mucoadhesive, which is potent over a wide pH range, from 4.5 to 8.5.[32] In contrast to most of other mucoadhesive polymers, the mucoadhesive property of DOPA does not involve hydrogen bonding. Since DOPA has shown strong adhesive properties towards a broad range of surfaces,[66,67] it remains to understand whether a specific interaction mechanism is involved in mucoadhesion. Li *et al.* have demonstrated a local modification technique for the AFM tip apex with mucoadhesive films, by means of microsyringe.[33] A localized coating of the AFM tip apex with PLGA resulted in a repulsive interaction with mucin layers, but further coating of the tip apex with chitosan showed attractive interactions under ambient conditions.

Ligand-receptor interactions

In the context of exploring adhesive properties of mucins, the specific interaction between lectins and carbohydrate moieties of mucins is another excellent example. The recognition of specific interactions between lectins and carbohydrates is a long standing, extensively studied topic in glycobiology,[68,69] and the recent application of scanning-probe techniques reveals a possibility for quantitative measurements of these interactions.[70,71] A first example is the bacteria-mucin interaction. Several studies have shown that adhesion is a prerequisite for the colonization of bacteria and a key mechanism for their probiotic activity, especially for lactic acid bacteria.[72,73] In this regard, Dague *et al.* have used AFM to probe *in vitro* interactions between *Lactococcus lactis* (*L. lactis*) and a PGM layer.[34] The PGM layer was prepared on a PS substrate (thickness of 3.4 nm). *L. lactis* cells were immobilized by spontaneous coating on top of a PEI layer adsorbed on an AFM tip. The presence of the PGM coating strongly reduced the bacterial adhesion force with respect to the bare substrate, which is consistent with the previous studies.[74] Both specific and non-specific interactions were observed to be involved in the interaction with PGM coatings. While the adhesive forces between PGM film and either *L. lactis*-probe or bare AFM probe were quite similar, a slightly higher percentage of specific interaction was observed for *L. lactis*-probe than for the bare AFM probe. The same group extended the topic to compare two different strains of *Lactococcus lactis*, namely IBB477 and MG1820.[35] In contrast to MG1820, from which only 5% of force curves can be assigned to specific interactions, for IBB477 strain, 20% of force curves could be assigned to specific interactions. In addition, the adhesive forces for IBB477 strain were nearly twice as large as those of MG1820. This observation was consistent with higher *in vivo* survivability of the IBB477 strain. Based upon blocking control experiments, it was also confirmed that O-glycans in PGM play a major role in interactions with *L. lactis*.

Sletmoen *et al.* have measured the interaction between a single-molecule carbohydrate (α-GalNAc (Tn-Antigen)) extracted from porcine submaxillary mucin (PSM) with soybean agglutinin (SBA).[37] It is important to note that the mucin sample used in this study was the O-glycosylation domain of PSM, possessing α-GalNAc residues (Tn-PSM) only, as opposed to the entire mucin molecules. Tn-PSM was covalently anchored to the AFM tips. SBA is a tetrameric lectin with one carbohydrate binding site per subunit.[75] As a control, the interaction of Tn-PSM-functionalized tips with ConA-functionalized surfaces was measured, but no adhesion was detected. The results indicate that binding and unbinding events are continuously occurring up to $2\,\mu$m, in consistent with the length of mucin chains, supporting a binding model in which lectin molecules "bind and jump" from one α-GalNAc residue to another along the polypeptide chain of Tn-PSM before the final dissociation happens. In a follow-up study, the authors have shown that Tn-Antigen interacts not only with lectins, but also other Tn-Antigens,[38] i.e. self-interaction between glycans may take place. Since the strength and lifetime vary significantly depending on the carbohydrate decoration pattern, it cannot be generalized, but the results in this work suggest that carbohydrate-carbohydrate interactions may be involved in aggregation of mucins at high concentrations. While the aforementioned two examples showed specific interactions in which the glycan parts of mucins are participating, the studies by Sulcheck *et al.* have shown the binding force measurements between the peptide components of mucin (MUC1) and a single-chain variable fragment (scFv) antibody selected from a scFv library screened against MUC1.[39,40]

A fundamental question that one can raise in the studies with isolated mucins is the relevance of these model systems to real biological systems, i.e. the question whether mucin monolayers or even a glycan domain can represent the real mucosal surface. In this context, Baos *et al.* have carried out a very interesting comparison, namely probing the distribution of sialic acids in both purified ocular mucin molecules and tear fluid.[36] The authors used two types of lectins, namely *Maackia amurensis* (MAA) and *Sambucus nigra* (SNA), in order to exploit specific interactions with α-2,3 and α-2,6 sialic acids, respectively, on ocular mucin or tear fluid. This study showed the predominance of α-2,3 over α-2,6 sialic acids in purified ocular mucins, but the reverse trend in tear fluid. This difference may suggest that the ocular surface gel has evolved to package mucins in such a way that only certain ligands are available at the ocular surface. Furthermore, this study validates the importance of studying mucus gels in parallel with purified mucin molecules, despite their less-well-defined compositional and structural features, as will be mentioned again below.

2.2.3. *Lubrication*

To date, all studies of the lubricating properties of molecular mucins have been carried out by employing commercially available samples, i.e. BSM (Sigma-Aldrich, AG) and PGM (Sigma-Aldrich and Orthana). This section is to be divided into two speed and lubrication regimes: the fluid-film regime and the boundary lubrication regime.

Fluid-film regime

Gassin *et al.* employed a tribopair composed of a compliant disk (PDMS) and a rigid ball (steel), and acquired Stribeck curves from PGM solution (up to 0.2%) over the speed range from 0.003 m/s to 2 m/s under 10 N load.[4] As is well known, under these contact conditions, the isoviscous-elastic lubrication mechanism (also known as "soft EHL"), can be readily activated.[76,77] In soft EHL regime, the pressure does not lead to a change (increase) in lubricant viscosity, but tribopairs deform to lower the contact pressure. MTM is a unique experimental approach, in which the rotations of both tribopair surfaces, ball and disk, are controlled independently, such that the ratio of sliding and rolling can be controlled all the way from pure

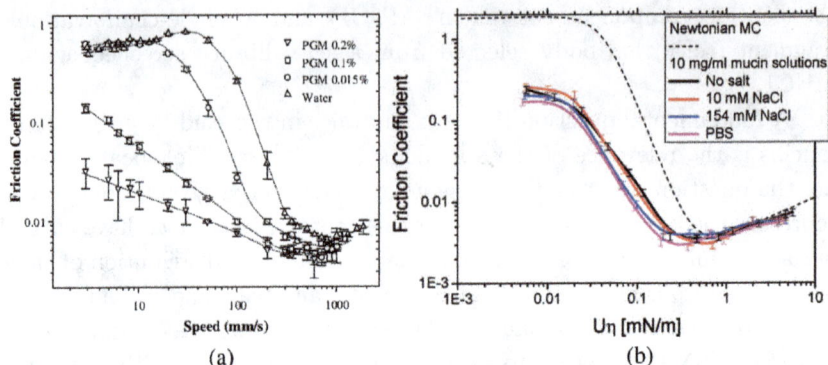

Fig. 2.6. (a) Stribeck curves obtained for water and various levels of PGM using untreated surface (from Ref. 4) (b) Stribeck curves at room temperature for various ionic strengths and types of electrolyte for 10 mg/mL mucin solutions (from Ref. 5).

sliding, mixed sliding/rolling, to pure rolling. With an increasing rolling component, the entrainment of lubricants into the contact inlet becomes more feasible. At an entrainment speed of 1 m/s, the central film thickness was estimated to be between 1 and 2 μm. A similar calculation for a 0.2% PGM solution, based on its Newtonian viscosity and its friction curves shown in Fig. 2.6(a), gave a film thickness value of 0.85 μm for entry into full hydrodynamic lubrication. No further detailed analysis of fluid-film lubrication was carried out.

Yakubov *et al.* have also characterized the lubricating properties of PGM ("Orthana") by means of MTM, but using PDMS for both ball and disk.[5] As shown in Fig. 2.6(b), the Stribeck curve, plotting friction coefficient as a function of the product of speed and viscosity (U · η), showed a fluid-film transition for 10 mg/ml PGM solution after U · η ≈ 0.4 mN/m. However, no further analysis on the high-speed regime data is provided in this work either. The lack of active research on the fluid-film regime is partly due to a lack of instrumental approaches to determine the film thickness under tribological contacts. While various interferometry techniques are available to determine the lubricant film thickness involving rigid reflective samples,[78] this is not readily applicable to soft materials. Bongaert *et al.* recently developed a new Raman-based MTM technique to enable the characterization of film thickness involving a soft tribopair, such as PDMS,[79] such that fluid-film lubrication studies based upon aqueous lubrication became possible. More studies in this research direction are expected in the future.

Boundary-lubrication regime

Apart from the two papers mentioned above, the majority lubrication studies of mucins to date have focused on their adsorption and boundary lubricating properties. A first AFM friction study on mucin was reported by Berg *et al.* by means of the colloidal-probe microscope.[9] Both colloidal probe (silica) and the substrate (silica) were hydrophilic surfaces. The main interest was to understand the molecular components responsible for the lubricity of the salivary pellicle among BSM, PRP-1, and statherin. The frictional properties of these components were studied in variation of concentrations up to 1 mg/ml. The results showed that PRP-1 is more effective in reducing the friction forces than BSM or statherin at this tribopair. Petterson *et al.* have also studied the nanotribological properties of BSM with AFM. Their particular interest was to study BSM and BSM + chitosan complex, and especially their interaction with surfactants, such as SDS, as already mentioned in a previous section.[8] When chitosan is adsorbed on the mucin layer, the friction force is dramatically increased compared to the pure mucin layer, the coefficient of friction reaching around 0.4. While the corresponding reduced repulsion of BSM + chitosan composite film was understood from the conformational change, i.e. compaction as already mentioned above,[8,43] increased friction of this layer is a more complex phenomenon, due to the fact that tribological shear can potentially create a more complex disturbance and restructuring of the film. For example, upon exposure of BSM + chitosan composite film to a SDS solution, lower frictional forces were observed than before exposure to SDS, but after rinsing the system, even higher friction forces than for the intact BSM + chitosan film were observed.

Harvey *et al.* have carried out nanotribological studies of "Orthana" PGM by SFA.[7] On both hydrophilic and hydrophobized mica surfaces, two friction regimes were observed.

On a hydrophilic mica surface (Fig. 2.7(a)), the low-friction regime reaches up to a load of about $40 \mu N$ (contact pressure ≈ 0.8 MPa), with $\mu \approx 0.02$ to 0.03. At higher loads, μ (slope) increased to a limiting value of 0.15 to 0.30. The low-friction regime at low loads on hydrophobized mica surface, however, extended to greater loads than on hydrophilic mica, namely about $60 \mu N$ (contact pressure ≈ 1.0 MPa) with $\mu \approx 0.01$ to 0.02, reflecting the more stable anchoring of PGM on hydrophobic surfaces. Low friction in the low-load regime was attributed to low interpenetration of opposing layers, with energy dissipation arising from the viscous rubbing of segments past each other. The hysteresis observed in friction forces on

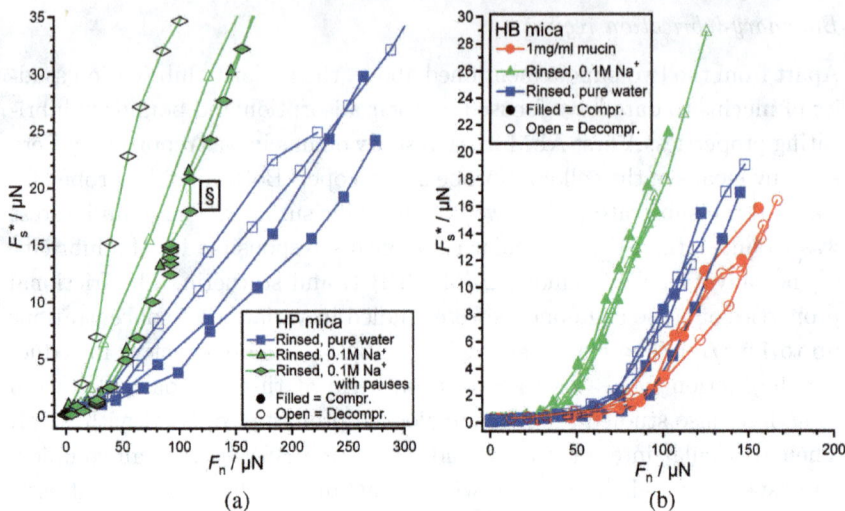

Fig. 2.7.   (a) Friction vs. load plots for "Orthana" PGM on hydrophilic mica surface.
(b) Friction vs. load plots for PGM on hydrophobized mica surface (from Ref. 7).

hydrophilic mica surface probably has similar origins to that in surface
forces, i.e. bridging of the two surfaces by mucin molecules, and leads to
non-linear friction responses. This bridging of the gap is a signature of
interactions between surfaces bearing adsorbed polymer layers, even under
good-solvent conditions[80] and even when the adsorbed chains are charged.[81]
Surface-anchored PGM layers are not as lubricious as other ideally stretched
and densely packed "brush-like" synthetic polymer chains, as assessed by
SFA.[82,83] This is because the packing and conformation of mucin molecules
on the surface is not ideally formulated on the surface, in contrast to the
case for synthetic polymer brushes that show extremely low friction forces.
In other words, a highly stretched, brush-like conformation of hydrophilic
chains (glycan chains for the case mucin) is not likely to be formed. The
authors argued that mucins adsorb in side-on configuration on a mica surface. Poor lubricating capabilities of PGM, as characterized by SFA have
been commonly observed for other biomacromolecules that are known to be
responsible for lubrication, such as lubricin[10,11] or hyaluronic acids.[84] This
is firstly because the *in vitro* conditions present in the SFA or other standard
friction-testing rigs are different from those of *in vivo* conditions. However,
it may also suggest that *in vivo* biolubrication is achieved synergistically
by several biomolecules, and thus isolated, single biomolecules cannot show
as effective a performance as that observed under *in vivo* conditions.

Lee *et al.* have carried out conventional pin-on-disk tribometry assessment of tribological properties of PGM (Sigma-Aldrich) at soft, elastic contacts by employing two PDMS surfaces as a tribopair.[6] In particular, this study focused on the change of conformation of PGM in bulk solution, adsorption onto the PDMS surface, as well as the frictional properties upon variation of the solvent environment, e.g. pH and ionic strength. This is important since the environmental conditions along the gastric mucosal surfaces, in particular pH, change significantly during digestion of swallowed foods. A major conclusion is that PGM is more aggregated and more slippery under acidic conditions rather than at neutral or basic pH, unless the ionic strength is too high (close to 1 M). Enhanced aggregation of PGM under low-pH conditions has been confirmed by means of many other techniques, including light scattering,[84] rheology,[86] and AFM.[87] It is also important to note that the higher adsorbed mass of mucins on the surface (at pH 7) is not directly translated into better lubricating properties.

2.3. Mucus and Mucosa

While isolated and purified mucin molecules offer opportunities to understand fundamental molecular behavior, many biophysical properties, such as viscosity or protective properties, can only be understood on the level of mucus gels. As is well known, the viscoelastic properties of mucus gels are not reproduced by highly concentrated mucin solutions.[90–92] Furthermore, the aforementioned example of probing specific carbohydrate moieties of purified ocular mucin vs. human tears has also revealed a significant difference between molecular mucin and mucus gel,[36] and indicates that purified molecular mucins do not necessarily represent physiological conditions. Some studies further emphasized the necessity to use not only mucus gels, but mucosal tissues. For example, Meyer *et al.* have emulated eye blinking with biological tissues as well as with synthetic materials as model tissues,[88,89] and concluded that metal-on-metal and tissue-on-metal systems were not predictive of the results obtained from the more clinically relevant tissue-on-tissue system for ophthalmologic purposes. In a related context, tests of the applicability of biomedical devices and implants that come into contact with mucosal tissues, e.g. catheters, endoscopes, and contact lenses, should be carried out under *in vivo* conditions — simple tests against molecular mucins are often not of relevance. Interfacial force studies of mucus and mucus/mucosa have been most active in two areas, saliva and lingual mucosa (2.3.1) and gastrointestinal tracts for the microcapsule-style endoscope (2.3.2).

2.3.1. *Saliva and lingual mucosa*

Ranca *et al.* have measured changes in the friction coefficient and wettability of oral mucosal tissue (piglet) induced by coating with a salivary layer (human).[93] The counter-surface (slider) for friction tests was a stainless-steel ball with 10 mm diameter, normal load being 0.1 N, and velocity 0.5 mms^{-1}. As expected, the average value of the dynamic friction coefficient of the "uncoated" tongues is reduced upon coating with saliva. This study is, of course, very simple, nevertheless is experimental proof that saliva indeed reduces the frictional forces of oral mucosal contacts. Berg *et al.* have also shown the lubricating effect of saliva at a silica/silica sliding interface by AFM.[94] The friction coefficient dropped from 0.66 to 0.03 across 1 mmol NaCl aqueous solution by placing a layer of saliva on the silica surface. The interfacial forces involving saliva have chiefly focused on slipperiness. Sub-topics regarding saliva lubrication include; (1) influence of the source and method of saliva collection on lubricating properties (2) the components of saliva that are responsible for lubrication (3) the tribological properties of saliva-food mixtures, and (4) wear resistance and integrity of saliva films.

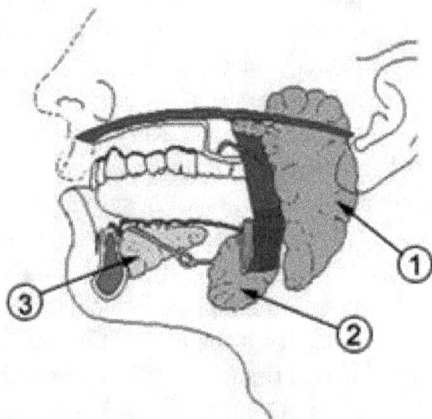

Fig. 2.8.   (1) parotid gland (2) submandibular gland (3) sublingual gland (Figure extracted from Ref. 96).

Source(s) and pattern of saliva collection

Rhee *et al.* have studied the frictional properties of saliva originating from different source(s), namely, human parotid gland vs. submandibular and sublingual glands.[95]

Due to the fairly narrow speed range, 1.99 mm/s to 7.84 mm/s, and high contact pressure, 3.8 N to 19.5 N at the bovine enamel/enamel interface, no significant variation in friction was observed as a function of these parameters. However, the saliva from submandibular and sublingual glands showed a viscosity of almost twice that of parotid saliva and water, and larger variations in friction (decreasing with increasing speed), and thus holds a greater potential for fluid-film lubrication at higher speed range. A related issue is the pattern of saliva collection, namely, stimulated vs. non-stimulated saliva. Prinz *et al.* have studied load dependency of the coefficient of friction of oral mucosa by measuring the friction forces between two mucosal surfaces (everted porcine esophagus vs. tongue) lubricated with human saliva and other lubricants.[97] In comparison of stimulated and unstimulated saliva, lubrication with unstimulated saliva resulted in a lower coefficient of friction than with stimulated saliva (Fig. 2.9).

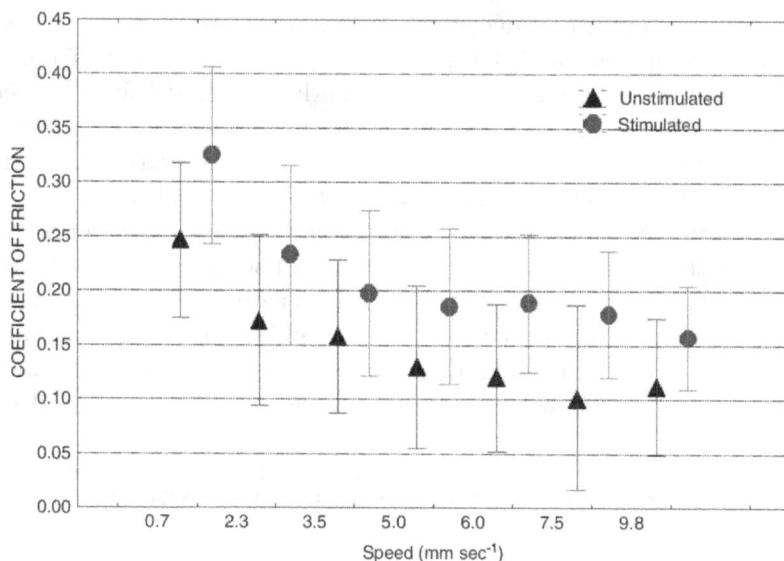

Fig. 2.9. Comparison of the coefficients of friction for mucosa, lubricated with stimulated and unstimulated saliva (From Ref. 97).

The authors related this difference to the difference in the source of saliva as mentioned above; stimulated saliva is derived mainly from the parotid gland and has a lower protein content and viscosity than unstimulated saliva. Unstimulated saliva is predominantly derived from the mucin-rich sublingual and submandibular glands, and the production is required only for eating.[98]

Components of saliva

Saliva contains several biomolecular species, and there has been a long debate over which components are primarily responsible for lubrication. In this regard, the most conventional view pinpoints saliva mucin (MUC5), and several studies have shown that the salivary mucins are chiefly responsible for the lubrication under light masticatory conditions.[98–102] These glycoproteins and their complexes provide thin-film lubrication under dental conditions of low occlusal load and high sliding speed. Successive studies have, however, revealed that other macromolecular species also contribute to the lubricating properties of saliva. For example, Douglas et al. focused on statherin (see the primary structure of statherin in Fig. 2.10) and showed that statherin shows a tendency to improve the boundary lubrication of oral mucosa.[103]

While the effect of reducing the friction forces by statherin is clear, it should be also noted that the absolute values of coefficient of friction under lubrication by statherin are still fairly high (> 0.4 for all speeds investigated). This is probably because heavy glycosylation is absent in the structure of statherin. Sample preparation in this work is relatively crude; the whole saliva was fractionated into several pools by size-gel filtration, and the fraction containing statherin was directly used for friction testing without further purification. Despite a risk of reduced purity, this approach offers the advantage of retaining the native conformation of proteins. Recently, Harvey et al. have further purified statherin molecules from parotid saliva and characterized the surface forces and friction forces by SFA.[104] By employing both hydrophilic and hydrophobized mica surfaces,

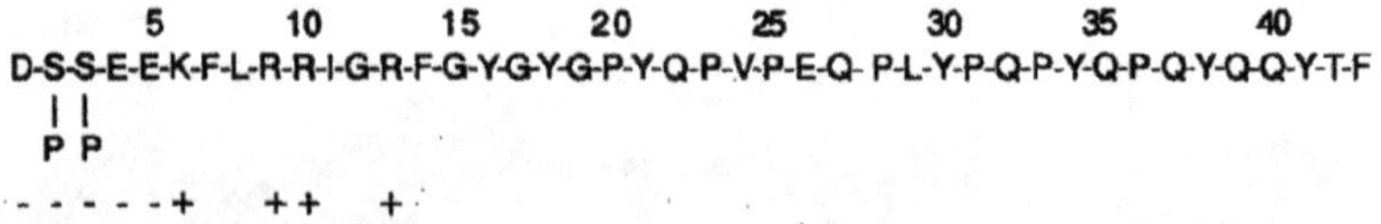

Fig. 2.10.   The primary structure (43 residues) of statherin (from Ref. 103).

the authors have shown much higher friction forces from statherin adsorbed on a hydrophobized mica surface ($\mu \approx 0.88$) than hydrophilic mica surfaces ($\mu \approx 0.09 - 0.12$). This difference was attributed to the amphiphilic nature of statherin, and formation of different multilayer structures on the two different mica surfaces. Hatton *et al.* have studied the lubricating properties of the PRG-albumin complex at a sliding pair of human cuspid tooth vs. glass plate. PRG is the major N-linked glycoprotein of serous parotid saliva.[105] The lubricating ability of the PRG of parotid saliva was observed to be enhanced by its interaction with human serum albumin. However, the interactive effect of albumin disappeared upon chemical deglycosylation of the glycoproteins, indicating that the interaction with albumin occurs with the carbohydrates of PRG. Fluorescence spectroscopy with a hydrophobic dye verified the existence of a PRG-albumin complex and demonstrated that deglycosylation of PRG altered the nature of its interaction with albumin. Bongaerts *et al.* have shown the contrasting influence on the rheological and boundary lubrication properties by isolating high-molecular-weight components from human whole saliva (HWS).[106] From "centrifuged saliva", for which whole saliva is centrifuged at 7.2 g force and only the supernatant phase was collected, the bulk rheological properties of HWS are altered significantly. Given that the viscosity and elasticity of fresh HWS arise mainly from the high-molecular-weight mucins, isolation of them by centrifugation, and in fact, aging of HWS for 24 hours in ambient too, almost completely removes the shear-thinning and elastic properties observed with fresh HWS. In contrast, the boundary-lubricating properties were hardly affected (see Fig. 2.11).

All these studies collectively indicate that the high-molecular-weight, supra-molecular mucins in saliva, are responsible for its rheological properties, but are not necessarily solely responsible for its boundary lubricating properties. Finally, as mentioned earlier, Berg *et al.* have studied the lubricating properties of three major components of saliva by AFM, and showed that all of them can effectively lubricate silicon substrates.[9] In this molecular-level study, PRP-1 turned out to be better lubricant than statherin or BSM at this interface.

Saliva-food mixtures

An advanced topic in this research area is to characterize the tribological properties of saliva–food mixtures. Zinoviadou *et al.* have characterized the tribological properties of saliva-food mixtures, in particular neutral

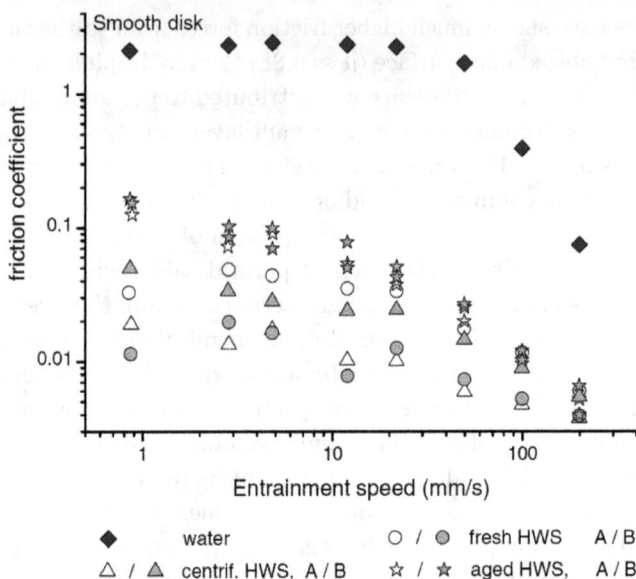

Fig. 2.11.   Speed-dependent friction coefficients for subjects A and B for fresh human whole saliva, centrifuged HWS and aged saliva (from Ref. 106).

polysaccharides, by means of MTM (neoprene ring vs. flat neoprene rubber).[107] This study aimed to assess the effect of mixing of saliva with two types of polysaccharide solutions, namely cross-linked starch from tapioca and locust bean gum (LBG), which represent carbohydrates that are affected (starch) and not affected (LBG) by enzymatic degradation (amylase) in the mouth, respectively. As expected, a strong decrease in viscosity upon addition of saliva was observed from the starch-containing sample. On the other hand, the addition of saliva did not affect the bulk rheological properties of LBG due to the lack of enzymes in saliva that can digest galactomannans. The variation in rheological properties was reflected in the friction forces; for LBG, addition of saliva did not affect the friction coefficients, whereas, for starch solution, addition of saliva resulted in increase of the friction forces, caused by enzymatic digestion. However, substantially lower frictional properties of starch solution compared to those of LBG persisted even after the treatment with saliva and enzymatic degradation. This suggests the presence of other components of the starch sample that contribute to its lubricating behavior, apart from the viscosity change caused from the digestion by saliva. The authors identified the presence of gelatinized granules in the starch sample, and proposed that remaining granules even after saliva-degradation are responsible for the low friction forces.

Wear resistance of saliva films

Stability and wear resistance of saliva coating are long-standing issues in odontology. Macakova *et al.* have studied the influence of ionic strength on the tribological properties of pre-adsorbed salivary films on compliant hydrophobic surfaces by means of MTM.[108] The adsorbed salivary film significantly decreases the boundary friction of PDMS surface due to the amphiphilic character of the adsorbed film and its structure. When bulk saliva is replaced by a protein-free salt solution, the pre-adsorbed proteinaceous salivary film becomes susceptible to a shear-induced wear at high loads when ionic strength is reduced from that corresponding to oral physiological conditions. At low loads (0.5 N), these structural changes did not lead to any deterioration of boundary lubrication. However, at high loads, the lubricious properties were dominated by the films' insufficient resistance to wear, which was exacerbated by a decrease in ionic strength.

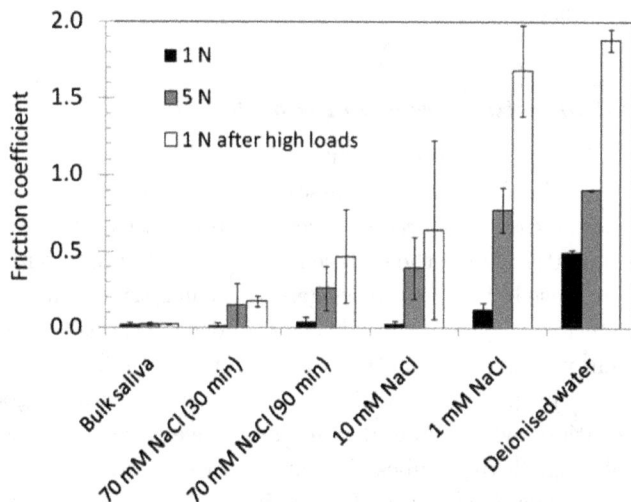

Fig. 2.12. Wear of a salivary film exposed to solutions with different ionic strengths. The values of friction coefficients at load 1 N obtained during the stepwise increase in load (black columns), at 5 N (gray columns) and after decrease in the load back to 1 N (transparent columns). The entrainment speed was 5mm/s. The error bars correspond to standard deviations between at least three independent experiments (from Ref. 108).

Several mechanisms may be responsible for this wear, including bridging of proteins adhering to opposing tribopairs, a loss of network structure of the adsorbed film, or desorption of the lubricious components from the adsorbed films. Applications of high loads led to a gradual loss of lubrication due to shear-induced wear of the films. Sotres *et al.* have also investigated the stability of salivary films against lateral shear, but on a nanoscale by means of AFM.[109] The authors have employed both hydrophilic and hydrophobic substrates, and compared the shear resistance at neutral and acidic pH conditions. The results can be characterized by two features; it is easier to remove saliva films from a hydrophobic substrate than from a hydrophilic one, and it is also easier to remove them at low pH than at neutral pH.

Finally, the influence of saliva viscosity on the tribological behavior of tooth enamel, both friction and wear properties, has been studied by Sajewicz.[110] The author gave the variation of viscosity of human saliva as being from 1.09 to 3.98 mP·s, as a function of aging. Tribological properties were characterized between two human premolar enamels. While the coefficient of friction data were very scattered when the viscosity was lower than 2.08 mPa·s, the corresponding wear properties (wear volume) were more clearly distinguished in saliva groups of low- and high-viscosity, with the threshold of 1.68 mPas.

2.3.2. *Gastroinestinal tract and capsule endoscopy*

Ikeuchi *et al.* have studied the lubricity of internal surface of the small intestine in a simple way.[111] In order to assess the role of the superficial layer of intestine mucosa, they measured the frictional properties of the intestine tissue before and after wiping the surface layer (soft and hard wiping). Wiping resulted in increase in friction forces against silicon nitride (flat), confirming that the superficial layer of the mucosa is indeed responsible for lubrication. As such, mucus layers on mucosa are conventionally associated with their "slippery" properties. Traditional tubing endoscopes are a good example of applications where lubricity on the surface is required; endoscopes are made of long, flexible tubes that are often covered with hydrophobic polymeric materials. Insertion and operation of endoscope usually requires minimization of the sliding friction between the device and the mucosa surfaces. For this reason, numerous methodologies to modify endoscope surfaces with slippery coatings have been developed.[112-114] Recently, it has been proposed that a microcapsule-style endoscope may provide a better diagnostic approach for endoscopic examination. As shown in the Fig. 2.13,

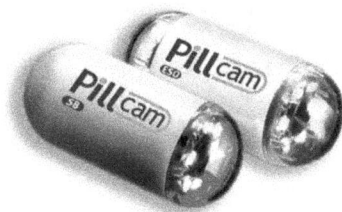

Fig. 2.13. Microcapsule endoscope, PillCam SB and PillCam ESO. The dimension is 11 mm × 31 mm. (from Ref. 115).

in this new approach, this "micro-robot" can be swallowed, travel along the digestive tract path, and the pictures sent wirelessly.[115] Most importantly, these new devices allow direct imaging of the entire small intestine, while conventional tubing endoscopes can only access the superior and inferior extremities.

In the practical operation of this device, a primary challenge is locomotion along the gastrointestinal tracts. The soft, viscoelastic, compliant, and varying morphology of mucosa along the digestive tract acts as a resistance for the device to move along. Moreover, the surface is slippery due to a mucus layer that covers the gastrointestinal mucosa, which varies in thickness and viscosity along the different tracts (Fig. 2.14).[116]

For full control of the locomotion of the devices, it should be possible to move forwards, stop, and even backwards, and thus the abilities to effectively grip on the mucosal surfaces are required. A most appealing and actively studied approach is to make use of adhesion as a basis for gripping at mucosal wall. Generating friction without applying high normal forces is a key requirement in order to eliminate the risk of tissue damage. Thus, instead of slipperiness, using stickiness of the mucosal layer has become a central issue; a number of approaches have been proposed to improve the adhesive properties against mucus gels, including mucoadhesive polymer films, geometrical parameters, and the combination of both.

Dodou *et al.* have employed Carbopol 971P NF (PAA-based) as mucoadhesive films.[117,118] The friction forces were measured between a Plexiglas surface covered with mucoadhesive films and the inside of a pig colon at a temperature of 37–38°C. Friction values in the presence of a mucoadhesive film were almost 20 times higher than those on the bare plate, showing the strong benefit of using mucoadhesive films to increase friction. Dodou *et al.* have also used geometrical variation as another approach to increase friction.[119] In this work, the dependence of the friction of

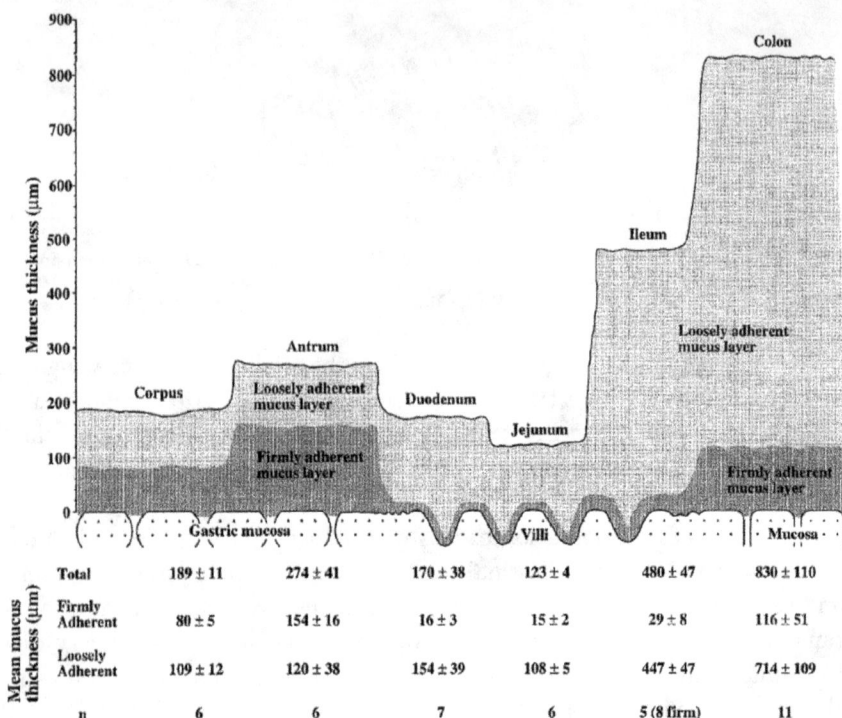

	Corpus	Antrum	Duodenum	Jejunum	Ileum	Colon
Total	189 ± 11	274 ± 41	170 ± 38	123 ± 4	480 ± 47	830 ± 110
Firmly Adherent	80 ± 5	154 ± 16	16 ± 3	15 ± 2	29 ± 8	116 ± 51
Loosely Adherent	109 ± 12	120 ± 38	154 ± 39	108 ± 5	447 ± 47	714 ± 109
n	6	6	7	6	5 (8 firm)	11

Fig. 2.14.   Mucus layer thickness changes along digestive organs (from Ref. 116).

mucoadhesive films on the geometrical nature of the sliding pad is identified, including features such as shape of pad, the presence of hole(s) and its
(their) position(s), and film orientation. By altering the film geometry, the
level of generated friction can be manipulated, and can be switched between
high and low values. Moreover, choosing geometries that can achieve high
friction can lead to a decrease in the overall size of the device. Efforts to
increase friction forces by optimizing the geometrical design or features
alone have also been reported by Y.-T. Kim et al. via multi-tube foot
design[120] and by S.-H. Lee et al. via micropatterning.[121]

The combined influence of mucoadhesive films and geometry has been
studied by Dodou et al.[122] Mucoadhesive micropatterns were prepared by a
double molding technique; an SU-8 lithographic structured wafer possessing
different patterned fields with pillars with a height of $20\,\mu$m, and diameters
and inter-pillar distances between 5 and $50\,\mu$m was used as master (see
Fig. 2.15(a)). Pig colonic mucosa was used as the counterface.

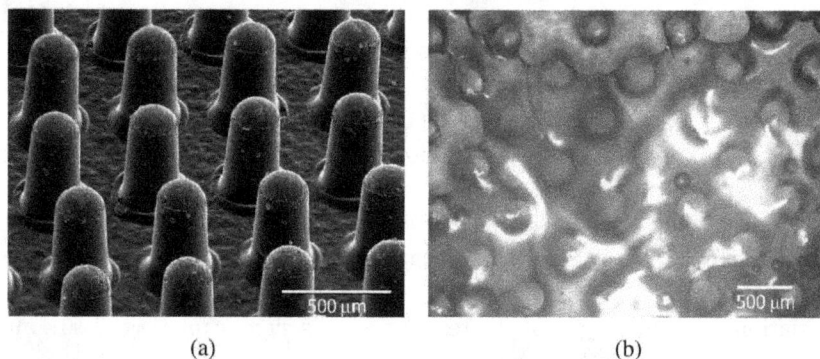

(a) (b)

Fig. 2.15. (a) SEM micrograph of non-mucoadhesive micropattern master (b) Surface of the epoxy-resin-cast non-mucoadhesive micropattern after shearing (from Ref. 122).

In vitro experiments showed that the grip and frictional performance of mucoadhesive + micropattern system exceeds that of non-patterned mucoadhesive films alone or micropatterns alone by ca. 1.5 times and 12 times, respectively. This can be attributed to a synergetic effect of the micro-patterned surfaces offering a more intimate contact with the mucus than flat surfaces. In addition, mucoadhesive pillars can create a non-continuous contact surface that inhibits the propagation of cracks through the mucus-mucoadhesive interface. It is interesting to note that static friction generated by micropatterns alone is far lower than static friction generated by mucoadhesives alone, indicating that the chemical approach is more effective than geometric manipulation in increasing friction forces. A distinct drawback of the micropattern approach, without mucoadhesive films, is that the frictional performance deteriorates with time; microscopic analysis of the patterns after experiments demonstrated that an increasing amount of mucus was entrapped in the inter-pillar gaps during the start-stop cycle (see Fig. 2.15(b)). Kwon *et al.* have also shown friction enhancement via combined manipulation of the micro-patterned and fluid layer, but instead of mucoadhesive films, they used a thin silicone oil layer on small intestinal surfaces.[123] The enhancement of the frictional force over a non-patterned flat elastomer material was roughly seven-fold.

Wang *et al.* have proposed porous bioceramic β-TCP as surface materials for microcapsules, due to its unique porosity and a potential to interact strongly with mucus layer when slid against intestinal mucus films.[124] Maximum static friction force was measured to vary with the surface pore size. Unfortunately, the test was not carried out with mucosa surface, but on a plate covered with ca. 200-μm-thick mucins prepared by dispersion, and thus a better proof of concept is needed.

Relatively scarce in this research area have been efforts to understand the mucosal side, i.e. the structural and biochemical features of mucosal surfaces and their influence on the friction against the device. In this context, the study by Terry *et al.* is a rare one that studied the position dependence of the friction coefficient of small bowels by classifying the proximal, middle, and distal regions, along the length.[125] While no significant difference in the friction coefficient was observed according to the regional origin, its variation over the length of the test segment is significant (Fig. 2.16), and this is attributed to the heterogeneity in the viscoelastic properties of the intestine.[126]

So far, all the examples reviewed for capsule-style endoscopes are based on passive locomotion, i.e. locomotion is entirely relying upon natural propulsion by peristalsis. For an active locomotion approach, both approaches increasing the adhesive grip and lowering friction forces for smooth gliding, have been reported. The work by Buselli *et al.* represents the former case;[127] the authors have fabricated the active capsule endoscope by means of bio-inspired polymeric micro-patterns, which are arrays

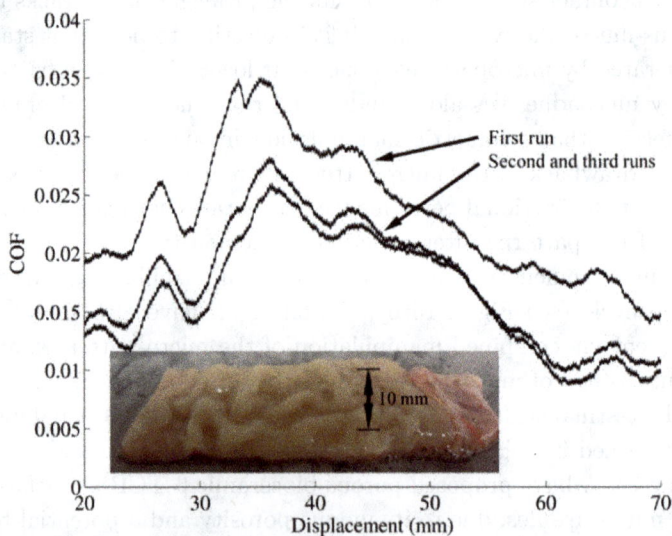

Fig. 2.16. COF versus sled position for three runs on a single sample (*in vitro* test, distal small bowel). Inset: portion of an intestine sample illustrating the relative size of mucosa topographical features (from Ref. 125).

of cylindrical pillars fabricated via soft lithography in order to enhance the grip between device and intestinal tissue. The patterns are mounted on micro-fabricated legs of a capsule robot that is able to move actively in the gastrointestinal tract, thus improving the robot's traction ability. The results showed that micro-patterns of pillars made from a soft polymer with an aspect ratio close to 1 enhanced friction by 41.7% with respect to flat surfaces. The patterned structure allows the soft intestinal tissue to conform to the polymeric surface. In addition, the mucus can fill the space in the grooves, improving the grip. Ciuti *et al.* have shown the possibility of exploiting external magnetic fields as a promising approach for the active guidance of endoscopic capsules.[128] The authors presented the development of a capsule prototype, which integrates permanent magnets with a vibrating motor and a triaxial accelerometer, along with an electronic module allowing remote control of the motor and wireless transmission of the inertial data to a host PC. In this case, reduction of the frictional forces is required to enhance the maneuverability of the device and reduce the risk of tissue damage during the examination. Reduction of the frictional forces occurs as the vibration motor frequency increases, allowing the movement of the capsule through small and large collapsed intestinal tissue.

2.3.3. *Other mucus and mucosa*

Studies focusing on interfacial forces on mucus other than saliva and gastrointestinal mucosa are extremely rare. One of the rare interfacial-force studies on mucus gels other than saliva or digestive mucosa is that of cervical mucus by AFM by Wan *et al.*[129] The authors have used AFM to characterize the mucin layer of various natural cancer cells, glycosylation-inhibited cells, and drug-resistant cells by measuring the mechanical force needed to penetrate the mucin layer and the associated mechanical energy, in the context of known drug resistance. An example is shown in Fig. 2.17 on the characterization of breast ZR-75-1 carcinoma.

In contrast to natural cancer cells, the glycosylation-inhibited samples do not possess the characteristic bumpiness in the normal forces prior to the cell deformation, because most sugar side branches on the mucin trunk are removed and the tip has virtually free access to the cell surface. Type-O-glycosylation is a cellular barrier known to reduce the impact of cancer drug therapies *in vivo*. This result is a simple force-distance curve that can be routinely obtained in an AFM laboratory, but its impact on

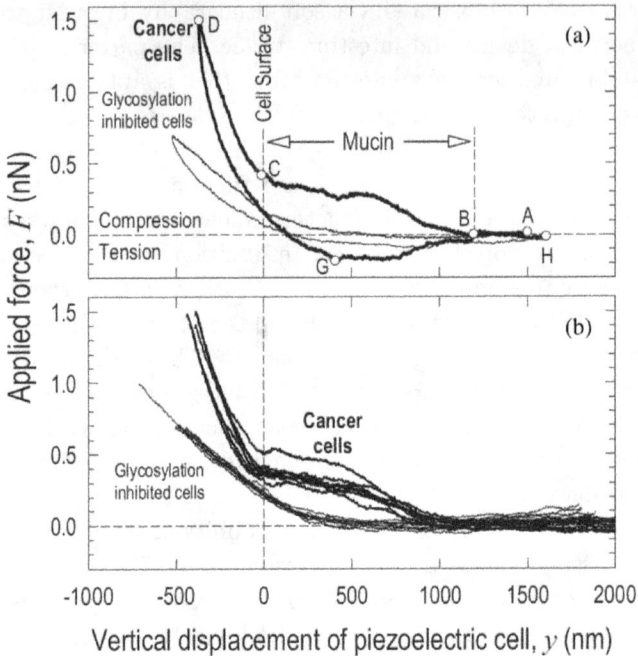

Fig. 2.17. (a) Typical mechanical response with and without glycosylation inhibition. Area bounded by curve BC and the horizontal axis is taken to be the mechanical energy needed to penetrate the mucin layer. (b) Ten loading curves for both natural and glycosylation-inhibited cells showing surprising reproducibility (from Ref. 129).

the visualization and understanding of the barrier role of mucin/mucus of cancer cells to drugs could be enormous.

2.4. Summary and Outlook

In this chapter, recent studies on the characterization and understanding of interfacial forces arising from interfaces involving mucins/mucus gels have been reviewed. Depending on the nature of the interacting counter-surfaces, the interfacial forces can be repulsive, adhesive, or slippery. Studies involving such interfacial forces measurements have been performed in the context of research in mucoadhesion, lectin-carbohydrate or protein-antibody binding, odontology, food science, mucosal endoscopy, biotribology, and have often involved mucosa, in order to provide relevance to physiological conditions. Unlike other biomolecules that are known to be responsible for the lubrication of biological systems, mucin/mucus gels have both adhesive and slippery characteristics, and both should be considered to fully understand the interfacial forces properties of mucin- or mucus-covered surfaces.

The major messages that can be extracted from the literature reviewed in this chapter can be summarized as follows. SFA, AFM, and tribometers have proven to be ideal approaches to study surface and frictional forces involving mucin layers. Mucin-mucin interactions appear to be repulsive by nature, as long as sufficient coverage is present on the surface, even though detailed interactions are significantly influenced by environmental conditions such as hydrophilicity of the substrate surfaces, pH, and the presence of salts in solution. The repulsive forces between mucin layers are primarily believed to be steric in origin. Due to the complex and hierarchical structure of mucin molecules, their conformation on a surface is not ideally stretched, as the case with certain synthetic, brush-forming polymer chains. This leads to fairly low friction forces, but not as low as has been reported for optimized, synthetic polymer brushes. Mucin layers show adhesive interactions towards many substances, including lectins, antibodies, and mucoadhesive polymer chains. SFA and AFM have proven to be valuable tools in mucoadhesion studies. Studies of the lubricating properties of mucins have mostly focused on the adsorption and boundary-lubricating properties, to date, in particular employing elastomers, in order to emulate the contact pressure experienced in biological tissues. Interfacial-force studies involving mucus gels have been mostly carried out for saliva and gastrointestinal mucus. Reflecting its multi-functional characteristics, a broad range of interfacial force properties have been studied for saliva, such as the influence of its origin and collection methods, components of saliva that are responsible for lubrication, tribological properties of saliva-food mixtures, and finally the wear resistance of saliva films. Capsule-type endoscopy has been used to pursue mucoadhesion as a viable means to increase grip capabilities without damaging mucosal surfaces.

Studies of the interfacial forces of mucins and mucus gels to date, however, reveal a few common problems. Firstly, resource of mucins is very limited. So far, BSM and PGM, especially those from commercial sources, are used in the majority of studies. This, in turn, results in limited physiological relevance for many applications. An example is the employment of gastric or submaxillary mucins for mucoadhesion studies exclusively targeting intestines. Secondly, even for the same type of mucin, molecular weight and structures are often significantly different. PGM studies by Harvey *et al.*[7] and Malmsten *et al.*[45] provided good examples as to how molecular weight (often determined by purification details) can result in significantly different surface behavior. BSM studies by Perez *et al.*[42] and Lundin *et al.*[44] also showed a significant difference in surface-force profiles, and the difference in molecular weight is likely to be partially responsible. Thirdly, even among available mucin samples, preparation of mucin layers is performed in very different ways, and the termination is generally not clearly characterized or identified. For example, the studies reviewed in this chapter have prepared the mucin layers by means of spontaneous adsorption onto hydrophobic surfaces,[4–7,34,45,46,48] spontaneous adsorption onto hydrophilic surfaces,[7–9,28,32,33,41–44,48] covalent bonding,[31,37,52,53] or spreading of mucin powders onto surfaces covered with epoxy glue.[30] Thus, the topmost layer of the prepared mucin films could be quite different in each case. This is a particular problem for studies focusing on specific interactions. Lastly, a holistic view of the different relevant length scales, i.e. those of mucin domains, mucins, mucin aggregates, and finally mucus on mucosal tissues, and the relationship between them, is lacking. Many studies reviewed in this chapter have also shown that a layer composed of isolated mucin molecules does not represent the biophysical properties of the mucus where the mucins originated. If we compare mucin with lubricin or glycoproteins from synovial fluids, mucus can be compared to the synovial fluid itself. While research on mucin molecules and mucus gels is widespread in several disciplines, efforts to link the individual results into a bigger picture will be valuable in the future.

2.5. Acknowledgements

The author greatly acknowledges support from European Research Council (ERC), Project Number 261152. The author would like to thank Professor Dr. Nicholas D. Spencer for his thoughtful advice and comments on the manuscript.

References

1. G. J. Strous and J. Dekker, *Crit. Rev Biochem Mol Biol* 27, 57 (1992).
2. R. Bansil, E. Stanely and J. T. LaMont, *Ann Review of Physiology* 57, 635 (1995).
3. R. Bansil and B. S. Turner, *Curr Opin Coll Interf Sci* 11, 164 (2006).
4. G. Gassin, E. Heinrich and H. A. Spikes, *Tribol Lett* 11, 95 (2001).
5. G. E. Yakubov, J. McColl, J. H. H. Bongerts and J. J. Ramsden, *Langmuir* 25, 2313 (2009).
6. S. Lee, M. Müller, K. Rezwan and N. D. Spencer, *Langmuir* 21, 8344 (2005).

7. N. M. Harvey, G. E. Yakubov, J. R. Stokes and J. Klein, *Biomacromolecules* 12, 1041 (2011).
8. T. Petterson and A. Dedinaite, *J Coll Interf Sci* 324, 246 (2008).
9. I. C. H. Berg, L. Lindh and T. Arnebrant, *Biofouling* 20, 65 (2004).
10. B. Zappone, G. W. Greene, E. Oroudjev, G. D. Jay and J. N. Israelachvili, *Langmuir* 24, 1495 (2008).
11. B. Zappone, M. Ruths, G. W. Greene, G. D. Jay and J. N. Israelachvili, *Biophys J* 92, 1693 (2007).
12. D. P. Chang, N. I. Abu-Lail, F. Guilak, G. D. Jay and S. Zauscher, *Langmuir* 24, 1183 (2008).
13. D. P. Chang, N. I. Abu-Lail, F. Guilak, G. D. Jay and S. Zauscher, *Soft Matter* 5, 3438 (2009).
14. L. Ng, H. H. Hung, A. Sprunt, S. Chunbiskaya, C. Ortiz and A. Grodzinsky, *J Biomech* 40, 1011 (2007).
15. S. Park, K. D. Costa and G. A. Ateshian, *J Biomech* 37, 1679 (2004).
16. L. Han, A. J. Grodzinksky and C. Ortiz, *Ann Rev Mat Res* 41, 133 (2011).
17. U. Meyer and H. P. Wiesmann, *Bone and Cartilage Engineering*, Springer (2006).
18. P. K. Singh and M. A. Hollingsworth, *Trends Cell Biol* 16, 467 (2006).
19. M. A. Hollingsworth and B. J. Swanson, *Nat Rev* 4, 45 (2004).
20. S. C. Chauhan, D. Kumar and M. Jaggi, *J Ovarian Res* 2, 21 (2009).
21. A. P. Moran, A. Gupta and L. Joshi, *Gut* 60, 1412 (2011).
22. C. Hattrup and S. J. Gendler, *Ann Rev Physiol* 70, 431 (2008).
23. T. Shirazi, R. J. Longman, A. P. Corfield and C. S. J. Probert, *Postgraduate Med J* 76, 473 (2000).
24. H. P. Hauber, S. C. Foley and Q. Hamid, *Canadian Resp J* 13, 327 (2006).
25. Y. Nakanuma, Y. Zen, K. Harada, H. Ikeda, Y. Sato, T. Uehara and M. Sasaki, *J Hepatobil Pancreatic Sci* 17, 211 (2010).
26. R. A. Cone, *Adv Drug Deliv Rev* 61, 75 (2009).
27. C. M. Evans and J. S. Koo, *Pharmacol Ther* 121, 332 (2009).
28. N. V. Efremova, Y. Huan, N. A. Peppas and D. E. Leckband, *Langmuir* 18, 836 (2002).
29. X. Zhu, J. DeGraaf, F. M. Winnik and D. Leckband, *Langmuir* 20, 10648 (2004).
30. J. Cleary, L. Bromberg and E. Magner, *Langmuir* 20, 9755 (2004).
31. M. Iijima, M. Yoshimura, T. Tsuchiya, M. Tsukada, H. Ichikawa, Y. Fukumori and H. Kamiya, *Langmuir* 24, 3987 (2008).
32. N. D. Carton, H. Lee and P. B. Messersmith, *Biointerphases* 1, 134 (2006).
33. D. Li, H. Yamamoto, H. Takeuchi and Y. Kawashima, *Euro J Pharma Biopharma* 75, 277 (2010).
34. E. Dague, D. T. L. Le, S. Zanna, P. Marcus, P. Loubiere and M. M. Mercier-Bonin, *Langmuir* 26, 11010 (2010).
35. D. T. L. Le, Y. Guerardel, P. Loubiere, M. Mercier-Bonin and E. Dague, *Biophys J* 101, 2843 (2011).
36. S. C. Baos, D. B. Phillips, L. Wildling, T. J. McMaster and M. Berry, *Biophys J* 102, 176 (2012).

37. M. Sletmoen, T. K. Dam, T. A. Gerken, B. T. Stokke and C. F. Brewer, *Biopolymers* 91, 719 (2009).
38. K. E. Haugstad, T. A. Gerken, B. Stokke, T. K. Dam, C. F. Brewer and M. Sletmoen, *Biomacromol* 13, 1400 (2012).
39. T. A. Sulchek, R. W. Friddle, K. Langry, E. Y. Lau, H. Albrecht, T. V. Ratto, S. J. De Nardo, M. E. Colvin and A. Noy, *Proceedings of the National Academy Sciences* 102, 16638 (2005).
40. T. Sulcheck, R. Friddle, T. Ratto, H. Abrecht, S. De Nardo and A. Noy, *Interdiscipl Trans Pheno* 1161, 74 (2009).
41. J. E. Proust, A. Baszkin, E. Perez and M. M. Boissonnade, *Coll Surf* 10, 43 (1984).
42. E. Perez and J. E. Proust, *J Coll Interf Sci* 118, 182 (1987).
43. A. Dedinaite, M. Lundin, L. Macakova and T. Auletta, *Langmuir* 21, 9502 (2005).
44. M. Lundin, T. Sanderberg, K. D. Caldwell and E. Blomberg, *J Coll Interf Sci* 336, 30 (2009).
45. M. Malmsten, E. Blomberg, P. Classon, I. Carlstedt and I. Ljusegren, *J Coll Interf Sci* 151, 579 (1992).
46. T. Pettersson, Z. Feldötö, P. M. Claesson and A. Dedinaite, *Surface and Interfacial Forces — From Fundamentals to, Book Series: Progress in Colloid and Polymer Science* 134, 1 (2008).
47. A. Dedinaite and M. J. Ernstsson, *J Phys Chem B* 107, 8181 (2003).
48. E. Perez and J. E. Roust, *Coll Surf* 9, 297 (1984).
49. M. P. Heuberger, M. R. Widmer, E. Zobeley, R. Glockshuber and N. D. Spencer, *Biomat* 26, 1165 (2005).
50. G. E. Yakubov, A. Papagiannopoulos, E. Rat, R. L. Easton and T. A. Waigh, *Biomacromol* 8, 3467 (2007).
51. T. Sandberg, H. Blom and K. D. Caldwell, *J Biomed Mat Res 91A*, 762 (2009).
52. M. Berry, T. J. McMaster, A. P. Corfield and M. J. Miles, *Biomacromol* 2, 498 (2001).
53. T. J. McMaster, M. Berry, A. P. Corfield and M. J. Miles, *Biophys J* 77, 533 (1999).
54. D. J. Brayshaw, M. Berry and T. J. McMaster, *Ultramicr* 100, 145 (2004).
55. N. A. Peppas and Y. Huang, *Adv Drug Deliv Rev* 56, 1675 (2004).
56. D. E. Chickering and E. J. J. Mathiowitz, *J Controlled Release* 34, 251 (1995).
57. S. A. Mortazavi and J. D. Smart, *Int J Pharmaceutics* 116, 223 (1995).
58. H. S. Ch'ng, H. K. Park and J. R. Robinson, *J Pharma Sci* 74, 399 (1985).
59. K. Park and J. R. Robinson, *Int J Pharmaceutics* 19, 107 (1984).
60. A. G. Mikos and N. A. Peppas, *S.T.P. Pharma* 2, 705 (1986).
61. L. Achar and N. A. Peppas, *J Controlled Release* 31, 271 (1994).
62. E. E. Hassan and J. M. Gallo, *Pharma Res* 7, 491 (1990).
63. F. E. Bailey Jr. and R. W. Callard, *J Appl Polym Sci* 1, 56 (1959).
64. R. D. Lunderberg, F. E. Bailey and R. W. Callard, *J Polym Sci* 4, 1563 (1966).

65. S.-J. Hwang, H. Park and K. Park, *Crit Rev Ther Drug Carrier Sys* 15, 243 (1998).
66. H. Lee, B. P. Lee and P. B. Messersmith, *Nature* 448, 338 (2007).
67. B. Lee, P. B. Messersmith, J. N. Israelachvili and J. H. Waite, *Ann Rev Mat Res* 41, 99 (2011).
68. M. Ambrosi, N. R. Cameron and B. G. Davis, *Organic & Biomol Chem* 3, 1593 (2005).
69. L. Kiessling and R. A. Splain, *Ann Rev Biochem* 79, 619 (2010).
70. E. L. Florin, V. T. Moy and H. E. Gaub, *Science* 264, 415 (1994).
71. P. Hinterdorfer and Y. F. Dufrene *Nat Meth* 3, 347 (2006).
72. C. Gusils, V. Morata and S. Gonzalez, *Meth Mol Biol* 268, 411 (2004).
73. A. C. Ouwehand, P. V. Kirjavainen, C. Shortt and S. J. Salminen, *Int Dairy J* 9, 43 (1999).
74. L. Shi and K. Caldwell, *J Colloid Interf Sci* 224, 372 (2000).
75. I. J. Goldstein and R. D. Poretz, in *The Lectins*; I. E. Liener, N. Sharon, I. J. Goldstein (Eds); Academic Press: New York, 35 (1986).
76. B. J. Hamrock and D. Dowson, *Proceedings of 5th Leeds-Lyon Symposium on Tribology on "Elastohydrodynamics and Related Topics"*, Mech. Eng. Pub. Bury St. Edmunds, Suffolk, 22 (1979).
77. M. Esfahanian and B. J. Hamrock, *Tribology Trans* 34, 628 (1991).
78. R. P. Glovnea, A. K. Forrest, A. V. Olver and H. A. Spikes, *Tribology Lett* 15, 217 (2003).
79. J. H. H. Bongaerts, J. P. R. Day, C. Marriott, P. D. A. Pudney and A.-M. Williamson, *J Appl Phys* 104, 014913 (2008).
80. J. Klein and P. F. Luckham, *Nature* 308, 836 (1984).
81. N. Kampf, U. Raviv and J. Klein, *Macromolecules* 37, 1134 (2004).
82. U. Raviv, S. Giasson, N. Kampf, J.-F. Gohy, R. Jerome and J. Klein, *Nature* 425, 163 (2003).
83. S. Lee and N. D. Spencer, *Achieving ultralow friction by aqueous, brush-assisted lubrication.* In, *Superlubricity*, Elsevier, 2007.
84. M. Benz, N. Chen and J. Israelachvili, *J Biomed Mater Res Part A* 71A, 6 (2004).
85. T. W. Waigh, A. Pagagiannoppoulos, A. Voice, R. Bansil, A. P. Unwin, C. D. Dewhurts, B. Turner and N. Afdhal, *Langmuir* 18, 7188 (2002).
86. J. Celli, B. Gregor, B. Turner, N. H. Afdhal, R. Bansil and S. Erramilli, *Biomacromol* 6, 1329 (2005).
87. Z. Hong, B. Chasan, R. Bansil, B. S. Turner, K. R. Bhaskar and N. H. Afdhal, *Biomacromol* 6, 3458 (2005).
88. A. E. Meyer, R. E. Baier, H. Chen and M. Chowhan, *J Adhesion* 82, 607 (2006).
89. A. E. Meyer, R. E. Baier, H. Chen and M. Chowhan, *J Biomed Mat Res Part B: Appl Biomat* 82, 74 (2007).
90. S. Rossi, M. C. Bonferoni, G. Lippoli, M. Bertoni, F. Ferrari, C. Caramella and U. Conte, *Biomaterials* 16, 1073 (1995).
91. J. Kocevar-Nared, J. IKristl and J. Smid-Korbar, *Biomaterials* 18, 677 (1997).

92. B. D. E. Raynal, T. E. Hardingham, D. J. Thornton and J. K. Sheehan, *Biochem J* 362, 289 (2002).
93. H. Ranc, A. Elkhyat, C. Servais, S. Mac-Mary, B. Launay and P. Humbert, *Colloids Surfaces A* 276, 155 (2006).
94. I. C. H. Berg, M. W. Rutland and T. Arnebrant, *Biofouling* 19, 365 (2003).
95. E. S. Reeh, A. Aguirre, R. L. Sakaguchi, J. D. Rudney, M. J. Levine and W. H. Douglas, *Clin Mat* 6, 151 (1990).
96. Wikipedia, "Salivary gland". http://en.wikipedia.org/wiki/Salivary_gland. Last accessed October 2013.
97. J. F. Prinz, R. A. de Wijka and L. Huntjens, *Food Hydrocolloids* 21, 402 (2007).
98. L. A. Tabak, M. J. Levine, I. D. Mandel and S. A. Ellison, *J Oral Pathol* 11, 1 (1982).
99. M. N. Hatton, M. J. Levine, J. E. Margarone and A. Aguirre, *J Oral and Maxillofacial Surg* 45, 496 (1987).
100. A. Aguirre, B. Mendoza, M. S. Reddy, E. A. Scannapieco, M. J. Levine and M. N. Hatton, *Dysphagia* 4, 95 (1989).
101. L. A. Tabak, *Crit Rev Oral Biology and Med* 1, 229 (1990).
102. A. M. Wu, G. Csako and A. Herp, *Mol Cell Biochem* 137, 39 (1994).
103. W. H. Douglas, E. S. Reeh, N. Ramasubbu, P. A. Raj, K. K. Bhandary and M. J. Levine, *Biochem Biophys Res Comm* 180, 91 (1991).
104. N. M. Harvey, G. H. Carpenter, G. B. Proctor and J. Klein, *Biofouling* 27, 823 (2011).
105. M. N. Hatton, R. E. Loomis, M. J. Levine and L. A. Tabak, *Biochem J* 230, 817 (1985).
106. J. H. H. Bongaerts, D. Rossetti and J. R. Stokes, *Tribol Lett* 27, 277 (2007).
107. K. G. Zinoviadou, A. M. Janssen and H. H. J. de Jongh, *J Food Sci* 73, E88 (2008).
108. L. Macakova, G. E. Yakubov, M. A. Plunkett and J. R. Stokes, *Tribol Int* 44, 956 (2011).
109. J. Sotres, L. Lindh and T. Arnebrant, *Langmuir* 27, 13692 (2011).
110. E. Sajewicz, *Tribol Int* 42, 327 (2009).
111. H. Yoshida, Y. Morita and K. Ikeuchi, *29th Leeds-Lyon Symposium on Tribology, Tribological Research and Design for Engineering Systems*, 425 (2003).
112. Y. Uyama, H. Tadokoro and Y. Ikada, *J Appl Polym Sci* 9, 489 (1990).
113. Y. Ikada, *Biomaterials* 15, 725 (1994).
114. D. D. Pothier, Z. Awad, M. Whitehouse and G. C. Porter, *Clini Otolaryngology* 30, 353 (2005).
115. G. Pan and L. Wang, *Gastroenterology Res Prac 2012*, 1 (2012).
116. C. Atuma, V. Strugala, A. Allen and L. Holm, *Amer J Gastrointestinal Liver Physiol* 280, G922 (2001).
117. D. Dodou, P. Breedveld and P. A. Wieringa, *Minimally Invasive Therapy* 14, 188 (2005).
118. D. Dodou, D. Girard, P. Breedveld and P. A. Wieringa, *Proceedings of the 12th International Conference on Advanced Robotics*, 352 (2005).

119. D. Dodou, P. Breedveld and P. A. Wieringa, *J Appl Phys* 100, 014904-1 (2006).
120. Y.-T. Kim and D.-E. Kim, *Proceedings of the Institute of Mechanical Engineering Part H: Journal of Engineering in Medicine* 223, 677 (2009).
121. S. H. Lee, Y. T. Kim, S. Yang, E. S. Yoon, D. E. Kim and K. Y. Suh, *Appl Mat Interf* 2, 1308 (2010).
122. D. Dodou, A. del Campo and E. Arzt, *Proceedings of the 29th Conference of the IEEE Engineering in Medicine and Biology Society*, 1457 (2007).
123. J. Kwon, E. Cheung, S. Park and M. Sitti, *Biomed Mat* 1, 216 (2006).
124. X.-Y. Wang, Y.-C. Han, X. Jiang, H.-L. Dai and S.-P. Li, *Biomed Mat* 1, 124 (2006).
125. B. S. Terry, A. B. Lyle, J. A. Schoen and M. E. Rentschler, *J Biomed Engrg* 133, 091010, (2011).
126. N. K. Baek, I. H. Sung and D. E. Kim, *Proceedings of the Institute of Mechanical Engineers Part H: Journal of Engineering in Medicine* 218, 193 (2004).
127. E. Buselli, V. Pensabene, P. Castrataro, P. Valdastri, A. Menciassi and P. Dario, *Measurement Science and Technology* 21, 105802 (2010).
128. G. Ciuti, N. Paternomichelakis, M. Sfakiotakis, P. Valdastri, A. Menciassi, D. P. Tsakiris and P. Dario, *Sensors and Actuators A*, in press (2012).
129. X. Wan, A. A. Shah, R. B. Campbell and K.-T. Wan, *Appl Phys Lett* 97, 263703-1 (2010).

Chapter 3

Aqueous Lubrication and Food Emulsions

Jason R. Stokes

School of Chemical Engineering
The University of Queensland, Brisbane
Queensland, Australia
jason.stokes@uq.edu.au

3.1. Introduction

Emulsions are an important class of aqueous-based lubricants as they can be designed to deliver essentially the same lubricating properties as oils under low-temperature situations ($<100°C$). Emulsions are utilised in many practical engineering applications including metal sheet rolling and ironing processes due to the ability of the water phase to dissipate heat and its non-flammability. Emulsions also have potential in low-temperature engineering applications as environmentally friendly lubricants using reduced amounts of oil and ingredients produced from renewable sources for environmental and economic benefits. Emulsion lubrication is also receiving attention in the context of food. Food products are often oil-in-water (o/w) emulsions and a typical food emulsion microstructure is shown in Fig. 3.1.[1] Studies into the lubrication properties of food emulsions are being used to develop insights into their behaviour during oral processing. Such advanced insights are leading to the ability to design healthier foods with acceptable sensory properties. In this chapter, we consider the general lubrication behaviour of o/w emulsions in traditional engineering applications as well as food emulsions in soft contacts, mimicking the conditions present during oral processing.

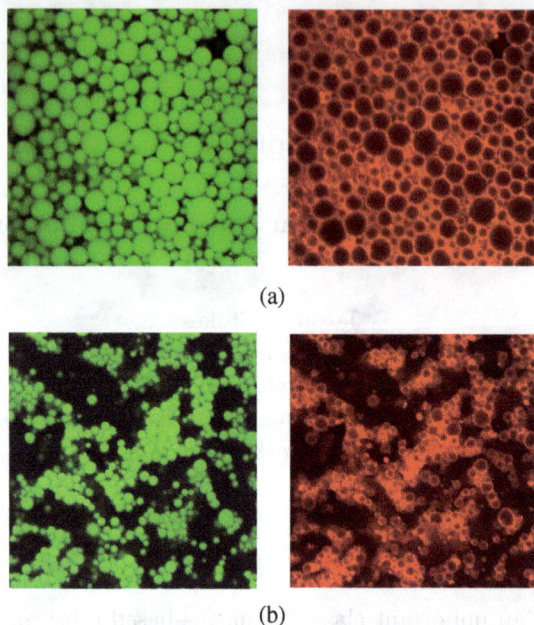

Fig. 3.1. Confocal image of a commercial mayonnaise emulsion (a) full-fat product, (b) low-fat product. Left image shows oil phase in green while the right image shows the protein stabilisers in red. Black regions in (b) are filled with hydrocolloid thickening agents such as starch. (Images reproduced from Ref. 1 with permission from the Royal Society of Chemistry.)

3.2. Emulsion Lubrication in Engineering

In an engineering context, utilising o/w emulsions for lubrication has several potential benefits due to the ability of the water phase to dissipate heat rapidly.[2] One of the primary applications for emulsions has been in metallic sheet rolling[3] and ironing processes[4] during the manufacture of thin metal sheets, cans and other metal objects. Emulsions are used to control the heat that arises due to the friction generated between tools and metal surfaces by preferentially entraining the oil phase into the contact zone for lubrication so that the surrounding water phase dissipates the heat.[2,5−8]

3.2.1. Friction and film thickness in rolling contacts

There has been considerable study of the behaviour of emulsions in hard tribological (e.g. metal-on-metal) contacts in the elastohydrodynamic (EHD) lubrication regime. Whetzel and Rodman[9] highlighted the potential of emulsions to be effective lubricants in cold-rolling of strip metal via the deposition of oil that was influenced by the use of a surface-active emulsifier. Advanced studies over the last few decades have utilised film thickness and visualisation of the contact zone to determine the properties of emulsions in rolling contacts, usually between a steel ball and flat transparent disc.

Optical interferometry has been used to measure the central film thickness of w/o and o/w emulsions and in many cases this is found to be generally close to that of the base oils. An example of an EHD rig for film thickness measurements is shown in Fig. 3.2, while Fig. 3.3 depicts the general behaviour of o/w emulsions in tribological contacts. Through control over emulsion stability and drop size, o/w emulsions can be designed to

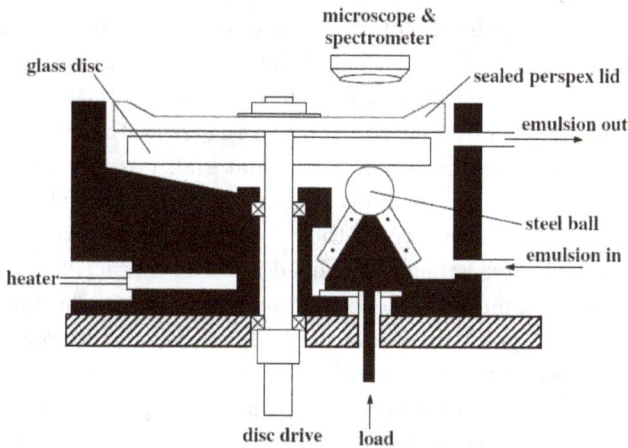

Fig. 3.2.   Schematic of an EHD film thickness rig modified for o/w emulsions. (Reprinted from Ref. 19 with kind permission from Springer Science+Business Media.)

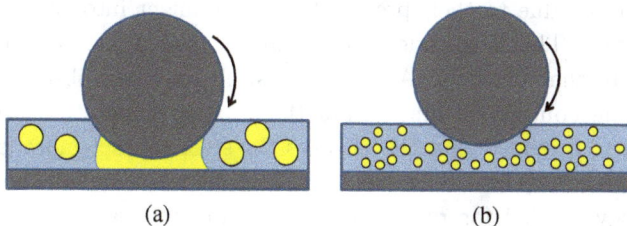

Fig. 3.3.   Schematic of o/w emulsions in tribological contacts, showing (a) oil pool formation from coalescence of large oil droplets between ball and plate, and (b) no oil pool formation from small stabilised oil droplets.

form a pool of oil in the tribological contact zone.[10,11] For stable emulsions with fine droplet sizes, Wang *et al.*[10,11] observes that the overall emulsion viscosity determined the film thickness, while a pool of oil forms in the contact when the emulsions is unstable. The process of oil pool formation also depends on entrainment speed; for example, Nakahara *et al.*[12,13] found that entrainment of oil to the contact is favoured at low speeds while at high speeds the overall emulsion governed the frictional response. Barker *et al.*[14] combined optical visualization and interferometry to correlate the size of the oil pool formed during EHD film formation using o/w emulsions. They showed that film thickness increases with increasing entrainment speed according to classical EHD theory based on the viscosity of the oil phase, until the local oil pool became so small that the distance between the inlet meniscus and the contact fell to such an extent that starvation ensued. The film thickness fell rapidly with further increase in speed in accordance with EHD starvation theory. It is also noted that at still higher speeds this trend is reversed and the film thickness starts to increase. This latter phenomenon is further investigated by Zhu *et al.*,[15] who suggested that it resulted from entrainment of a mixed oil/water phase material, possibly resulting from shear-induced coalescence of oil droplets in the inlet. Several other studies have also found that at low speeds the oil phase dominates the friction, but at higher speeds water is being entrained into the contact.[16–18]

Drop size has a major influence on the behaviour of emulsions in tribological contacts.[20] In particular, larger droplets favour oil-pool formation, which may arise due to the way different size oil droplets segregate in the flow. From hydrodynamics, it is expected that larger particles may segregate closer to the roll surface and are in forward flowing regions, while smaller droplets are closer to the flow centreline and in back flowing regions. Wilson *et al.*[21] show that for droplets larger than the lubricating film thickness, a 'concentration region' exists at the inlet where oil droplets become more densely packed due to their preferential entrainment into the inlet region of the contact. This is considered to occur because of oil's higher viscosity compared to that of water. At low speeds, oil has been observed to form a lubricating continuous film within the rubbing contact, referred to as 'plating out', provided that the oil wets the steel contact more favourably than the aqueous phase. This has been described in terms of the displacement energy needed for removing water from the steel and replacing it with the oil.[22] At higher speeds, the contact can become starved due to a shortage of supply of oil into the contact. This results in a drastic decrease in EHD film thickness above a critical speed, whereby the oil is replaced by a water-continuous phase. This critical speed depends on oil viscosity, surfactant type and oil phase volume, and aspects of the experimental observations have also been predicted theoretically.[4,6] A general finding is that the formation of a pool of oil in the inlet is governed by a balance between oil pool replenishment 'through instability' and loss 'through displacement of oil by the surrounding water phase, localized shear and by passage through the contact'.[10,11]

3.2.2. *General model of emulsion lubrication*

A significant body of work has been performed on emulsion lubrication in engineering contacts and a general picture has emerged on how emulsions lubricate, primarily, rolling EHD contacts. Wilson *et al.*[21] developed a dynamic-concentration model and later improved this using a mixed-flow model. The mixed-flow model, shown in Fig. 3.4, divides emulsion behaviour into three regions: supply region, concentration region, and a pressurisation region. The supply region is characteristic of the bulk o/w emulsion. As the emulsion moves through the supply region, the film thickness is such that the oil droplet can bridge the gap between the surfaces and become attached to, or wet, the surfaces. At this point the emulsion enters the concentration zone. The effectiveness of this droplet 'capture' process is represented by a capture coefficient (C) and the model assumes that the oil concentrates in the contact when the film thickness (h_i) is some

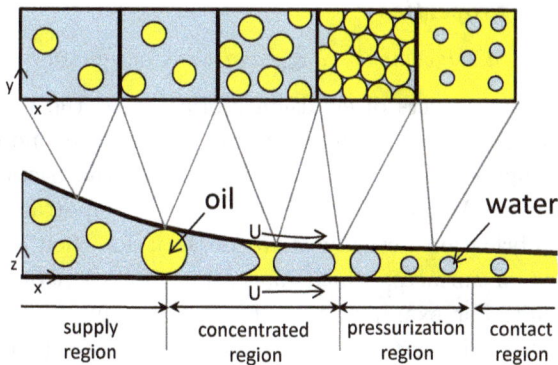

Fig. 3.4. Pictorial representation (drawn not to scale) of the mixed flow model for emulsion lubrication showing the oil phase from an o/w emulsion being concentrated between shearing surfaces. (Adapted from Ref. 21 with permission from Taylor & Francis Group, http://www.informaworld.com.)

fraction of the mean droplet size, d_s: $h_i = Cd_s$. At some point within this concentration regime, the emulsion inverts as the effective phase volume of oil increases (e.g. Wilson *et al.*[21] assumed an oil phase volume of 0.9 for inversion, although it was noted that inversion would be expected at lower phase volumes). In the pressurised regime, the lubrication behaviour is dominated by the oil continuous phase and the pressure dependence of the oil viscosity is an important defining parameter. When the rolling speed is increased, the lubrication properties are dominated by the oil phase that provides a reservoir of oil large enough to form a fully flooded, elastohydrodynamic contact. Thus the film thickness increases with increasing speed, but at some critical point, not enough oil is supplied into the contact and lubrication is dominated by the bulk emulsion or water phase.

Other models in the literature have been developed that essentially follow the approach of Wilson *et al.*,[21] but these models also contain a capture coefficient or similar variable.[7,23,24] Despite numerous studies into emulsion lubrication, and emergence of a general picture on the behaviour of emulsions, considerable research is still needed to relate emulsion lubrication to measureable properties of the emulsion, the surface tension at the solid-liquid interface, as well as determining the influence of other additives that are typically included in formulations. For example, it may be possible to relate the capture coefficient to physical-chemical parameters associated with the emulsion properties, surfaces and surfactant, and it is in this domain that future studies should focus.

3.2.3. *Surface properties*

3.2.3.1. *Wetting, displacement energy and work of adhesion*

One of the crucial aspects of emulsion lubrication is the wetting characteristics of the surfaces; the fluid that wets the surfaces more favourably is usually anticipated to be the fluid that defines the friction coefficient and film thickness. While wetting is regarded as important, few studies on emulsion lubrication actually characterise it for their system or use matching pairs of surfaces, which makes interpretation of much of this research difficult. For example, several comprehensive studies that characterise the film thickness use chromium-coated glass (silica) disk and steel ball in a rolling contact, whereby their surface chemistries are not matched.[2,19,22,25]

Wetting characteristics of the surfaces are important as both the surfaces, and the shear field they generate, drag the fluid between the ball and disk tribopair. Wetting is typically assessed by the individual surface tension of oil (γ_{OS}) or water (γ_{WS}) droplet on a solid surface exposed to air. This method has led to the favoured wetting phase in the emulsion to be evaluated using the concept of displacement energy, DE:

$$DE = \gamma_{OS} - \gamma_{WS} \tag{1}$$

To evaluate this parameter more easily, the contact angle of the oil and water can be used to characterise the degree of wetting; the liquid-solid contact angle (θ_{LS}) is the angle of the tangent between the droplet surface and solid substrate. For equilibrium at an air / liquid /solid interface, Young's equation gives:

$$\gamma_{LS} = \gamma_{SA} - \gamma_{LA} \cos \theta_{LS} \tag{2}$$

γ_{LA} is the surface tension at the liquid-air interface. Substituting into the equation for DE gives:

$$DE = \gamma_{LA} \cos \theta_{wS} - \gamma_{OA} \cos \theta_{OS} \tag{3}$$

DE reflects the relative work of adhesion to the surfaces, and if DE < 0, oil is expected to displace water, while the contrary is expected to happen if DE > 0. While several studies[2,19,22,25] state that full film lubrication followed that expected based on DE, it is apparent that this is not strictly the case. DE is still found to be positive under most circumstances, although trends are typically observed whereby a decreasing DE corresponds to a drive for oil to fill the contact. Cambiella *et al.*[19] also found that it is inadequate for predicting the behaviour of emulsions stabilised with anionic emulsifier. A clear explanation for this discrepancy is that the real lubrication situation involves wetting of the surface in the presence of both liquid phases rather than air. Rather than DE, the real driving force for adhesional wetting is given by the true work of adhesion that includes consideration of the interface between the oil and water, and is defined as:

$$W_A = \gamma_{OW} - \gamma_{OS} + \gamma_{WS}$$

The work of adhesion is the reversible work required to separate unit area of liquid from a substrate. Using Young's equation, the work of adhesion can be obtained from easily measurable parameters of the contact angle of oil at the solid-water interface (θ_{OWS}, as shown in Fig. 3.5) and the oil-water

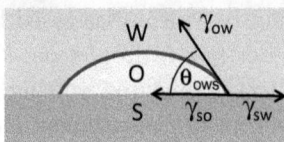

Fig. 3.5.   Three-phase contact angle (θ_{ows}) and surface energies for an oil droplet (O) on a solid substrate (S) in an aqueous fluid (W).

interfacial tension (γ_{OW}):

$$W_A = \gamma_{OW}(\cos\theta_{OWS} + 1)$$

The higher W_A, the stronger the adsorption of the oil film at the metal surface. Cambiella *et al.*[19] found that the work of adhesion gave a better indication of the film-forming ability of emulsions in rolling tribological contacts. The importance of measuring the 3-phase contact angle is apparent when one considers that the contact angle of oil at the steel/air interface is 26° compared with about 130° for oil at the water/solid interface[19]; hence, the presence of water decreases the ability of oil to wet the surfaces since water has a preference for the steel surface, which is more hydrophilic.

Persson *et al.*[26] also show the importance of characterising θ_{OWS} and highlight an alternative approach by using a spreading coefficient ($S_{L/S}$). This accounts for when a fluid in contact with a solid surface is displaced by another fluid that spreads over the solid surface[26]:

$$S_{L/S} = \gamma_{OW}(\cos\theta_{OWS} - 1)$$

Spreading will occur spontaneously if $S_{L/S} > 0$, while it is negative for partial wetting.

Several studies suggest that oil pool formation in EHD contacts is favoured by operating at surfactant concentrations about an order of magnitude below the critical micelle concentration (cmc) of the surfactant. This low surfactant concentration allows a self assembled surfactant monolayer to form at the steel surface, which renders the surface hydrophobic so that oil wets the surface at a minimum DE.[2,14,19,25] It should be noted that in these studies DE > 0, and yet oil still pools in the contact in all situations at low speeds. As shown in Fig. 3.6, there is a critical speed at which oil no longer pools in the contact, which presumably corresponds to a transition region towards where the water-continuous emulsion phase dominates the tribological behaviour. DE could not strictly be correlated to this point,

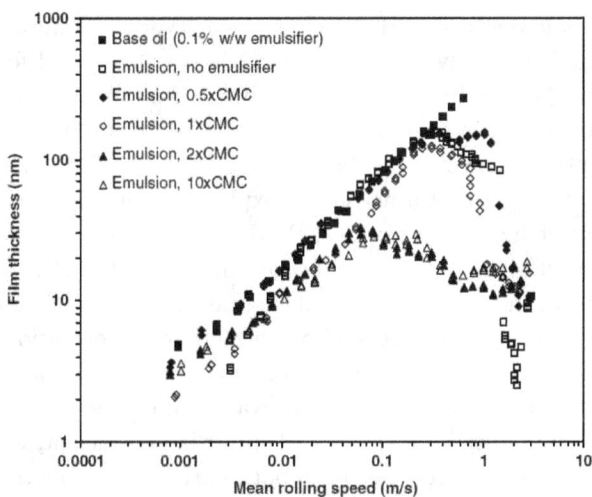

(a)

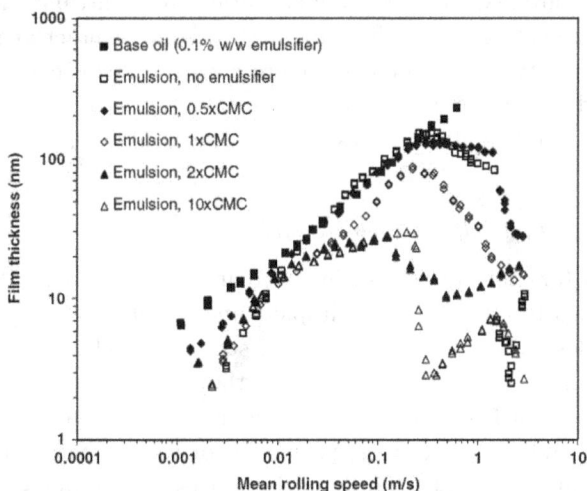

(b)

Fig. 3.6. Influence of emulsifier concentration (as function of the critical micelle concentration, cmc) on film formation for emulsions with (a) non-ionic and (b) cationic emulsifiers at 25°C in a rolling EHD contact. All emulsions had a base oil content of 3% w/w. Reprinted from Ref. 19 with kind permission from Springer Science+Business Media.

although it is hypothesised that it is related. Work of adhesion measurements were all relatively low, with the highest value found for emulsions without surfactant and the anionic surfactant; the OWS contact angle was $\gg 90°$ in most cases, indicating that the surfaces favoured the water phase at all concentrations, yet the tribological study does not reflect this.

Further research is still needed to explore the influence of surfactants and surface wetting characteristics. While it is very clear that emulsifiers have a major effect on emulsion lubrication, it is still unclear what drives the formation of the oil pool and the destabilisation of the oil pool at critical speeds. While these factors depend on surfactant concentration, this could either be due to the role of surfactant on emulsion stability against droplet breakup and coalescence or due to surface adsorption and interaction forces. It should be emphasised that the use of spreading coefficient, work of adhesion or displacement energy assumes both surfaces have matched wetting characteristics, which is not the case in most studies on emulsion lubrication. They are also measured under static conditions, but lubrication is a dynamic process so that advancing and receding contact angles may be more appropriate for developing a more accurate assessment of the film-forming and lubrication properties of emulsions. In addition, the balance of viscous shear and capillary forces has not been evaluated.

3.2.3.2. *Surfactant self assembly at surfaces*

Surfactants (or emulsifiers) are included in emulsions to both lower the energy needed to form emulsion droplets and stabilise the dispersed phase against coalescence. The very nature of surfactants also mean they readily adsorb and self assemble at surfaces; they can thus play a crucial role in the lubrication behaviour of emulsions and there is a need to balance stability[9,27] and surface-coating properties of the surfactant molecules.[25] Surfactant adsorption at solid-liquid interfaces is a complex subject and understanding the complexity of it remains a major challenge in colloid science. The kinetics, morphology, structure, and mechanisms of adsorption by cationic, anionic, non-ionic, and zwitterionic amphiphiles at the liquid–solid interface are still emerging.[28] Various studies show that, depending on the aliphatic chain length, aqueous electrolyte, concentration and surface chemistry of the substrate, surfactants may adsorb as hemicylindrical structures, cylindrical aggregates, monolayers, or bilayers. An example is shown in Fig. 3.7 for non-ionic surfactants. These assemble as a monolayer at very hydrophobic surfaces, while bilayer and micellular layers are observed with

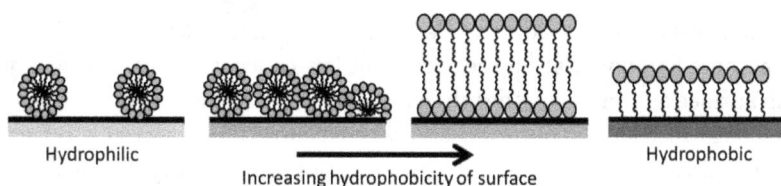

Hydrophilic ──────────────→ Hydrophobic

Increasing hydrophobicity of surface

Fig. 3.7. An example of the non-ionic surfactant coverage on surfaces that vary from hydrophilic to hydrophobic. Adapted from Ref. 30.

decreasing hydrophobicity.[29−31] It should be noted that the hydrophilic head group is outward facing in all these circumstances, indicating the surface is effectively rendered hydrophilic if the structure persists under non-static and tribological conditions.

In considering metal surfaces present in engineering contacts, it is important to note that the presence of charged groups on the surface and/or surfactant creates even more complex interaction between them; both are also dependent on pH. For example, oxide groups are present on metal surfaces that can undergo hydrolysis.[2,25] pH-dependent dissociation of surface-hydroxyl groups produces a positively charged surface at low pH and negatively charged surface at high pH. This electrical surface charge is neutralised by counter ions in the solution media to form the electrical double layer. The point of zero charge for steel is in the range 5.2–6.7 due to the presence of iron oxide species at the surface. The organisation of the surfactant at the surface depends on the charged groups present on the molecule, which depends on solution pH, as well as its molecular weight and concentration. Ratoi-Salagean *et al.*[2] suggest that their anionic surfactant adsorbs to hydrophilic/polar steel surface as a monolayer well below the cmc, with the negatively charged groups in the hydrophilic head group associated with the positively charged groups on the surface. As the concentration is increased, the surfactant assembles from the water phase at the steel surface to render it more hydrophobic. Hydrophobicity is a maximum at just below the cmc, as indicated by the contact angle at the water-steel interface. The contact angle then decreases with further increases in surfactant concentration that is associated with the formation of a bilayer of 'hemi-micelles' where the hydrophilic headgroups are exposed to the bulk liquid. Hence, the displacement energy is also dependent on both pH and surfactant concentration, and thus so is the lubricating properties of the emulsion. However, no evidence has been provided for the presence of a monolayer or bilayer of surfactant molecules.[19] While acknowledging that it is qualitatively correct, Persson *et al.*[26] noted that in reality "the surfactants lay more flat on the surface at low concentrations, and may well form surface aggregates rather than a double layer at higher concentration." They suggest that the observed critical concentration observed is reflective of the critical surface aggregate concentration that usually lies just below the cmc.

3.3. Emulsion Lubrication in Soft-Tribology and Food Applications

Lubrication has long been considered to be an important factor in the oral processing of foods,[32,33] although it is only recently that researchers have begun to try to evaluate such properties. There are surprisingly few studies published that actually consider the lubrication properties of full food formulations, despite the work of Kokini *et al.* (1977)[33] demonstrating that even simple friction measurements had the potential to correlate to food sensory properties such as 'smoothness' and 'slipperiness'. Two products that have been examined are chocolate and mayonnaise, but only limited conclusions could be drawn. The lubrication of molten chocolate is found to depend on fat particle behaviour at the inlet region of a sliding contact,[34] fat constitution and average particle size. When mayonnaise samples are confined to narrow gaps in a surface force apparatus, the response also depended on particle properties such as size, stiffness and hydrophobicity, and differences are observed between full-fat and low-fat products (Giasson *et al.*, 1997). However, it is difficult to interpret the tribological response of food products due to their innate complexity in terms of rheology and microstructure, while they also contain multiple phases and multiple components that are interfacially active.[1] This is in contrast to many of the aforementioned lubricants used within engineering environments, although even these are growing in complexity as greater degrees of functionality are built into them.

A major difference to the majority of studies into emulsion lubrication is that oral lubrication involves shearing against deformable rough substrates. This means that the shearing surfaces themselves are compliant (i.e. soft tribology) and that boundary and mixed lubrication regimes are likely to be important[35]; most of the aforementioned studies in emulsion lubrication consider primarily full-film lubrication under EHD conditions. In situations involving at least one 'soft' biological surface, the lubricant's rheology is not influenced by the pressure in the contact.[36,37]

The reason to study the lubrication of foods, and food emulsions in particular, is that it is anticipated that their properties in tribological contacts will be reflective of their behaviour during oral processing, which will subsequently affect momentum, heat and mass transfer. For example, it is anticipated that the heat- and mass-transfer of components that elicit a tactile, taste or aromatic sensation will be strongly dependent on the material that is present in the contact zone between rubbing surfaces. Therefore, having control over the behaviour of foods in the tribological contact between tongue and palate may lead to the ability to engineer foods with enhanced organoleptic properties and functionality, as well as the ability to obtain an acceptable consumer response using healthier formulations (e.g. reduced fat products). Recent approaches have been focused on understanding the tribological properties of model foods using o/w emulsions.

3.3.1. *Oral tribology*

Human tribological processes usually involve at least one deformable substrate (e.g. tongue), and so tribological studies fall under the realm of isoviscous elastohydrodynamic lubrication whereby the pressure in the rubbing contact deforms the surface rather than influencing the fluid rheology.[38] There are two general approaches to evaluate the role of tribology of food systems: (1) utilisation of a traditional ball-on-disk tribometer that is modified with deformable surfaces and operated over a large range of speeds and loads; (2) utilisation of sliding devices at limited speeds and loads, including the use of potentially realistic oral substrates (e.g. *ex vivo* pigs tongue). Two tribometers that have been used in the study of food emulsions are shown in Fig. 3.8.

The utilisation of modern tribometers provides a fundamental insight into the mechanisms by which the friction coefficient is defined in a soft-tribological contact for a complex food or emulsion system, particularly when the system is well defined in terms of surface roughness and surface chemistry. It should be noted that foods are structurally and rheologically complex, so any measured frictional response may arise from a variety of sources. However, most studies in soft tribology have focused on friction measurements rather than film thickness, which has been the focus of hard-contact emulsion lubrication. The focus on friction measurements arise because (a) film thickness measurement in soft contacts is not routine and methods are only now being realised,[40-43] and (b) the boundary and mixed regimes are more important in oral processes because of the roughness of

J. R. Stokes

(a)

(b)

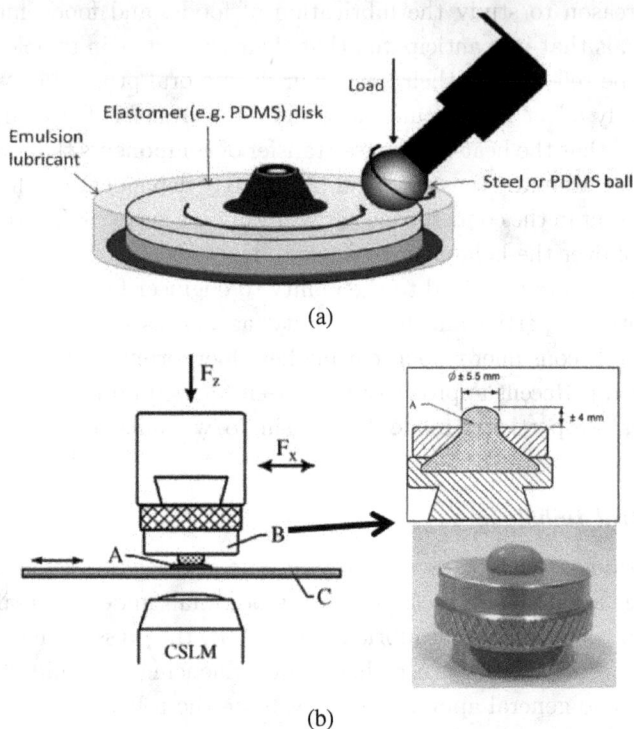

Fig. 3.8. Schematic representation of tribometers using in the study of food emulsion lubrication. (a) mini-traction machine, modified to utlise elastomeric surfaces such as polydimethylsiloxane (PDMS) in place of traditional steel tribopairs.[35] (b) Novel optical tribological configuration, showing the measurement probe (B) with a tongue sample screwed into the probe; Reprinted from Ref. 39 with permission from Elsevier.

oral surfaces. Without film-thickness measurements, the shear rate and thus the viscosity of the lubricant is ill-defined in the tribological contact. However, de Vicente *et al.*[37] and Bongaerts *et al.*[35] highlight that system master curves can be measured using the mini-traction machine (Fig. 3.8(a)), which covers the three lubrication regimes for particular tribopairs using aqueous Newtonian lubricants of varying viscosity; these can be used to determine the effective viscosity of the fluid in the contact zone. It is observed that the friction coefficient in the main portion of the hydrodynamic regime did not depend on the surface properties, even when the wetting and roughness of the surfaces are altered as shown in Fig. 3.9.

Oil is generally regarded as a good lubricant because of its relatively high viscosity and high pressure-viscosity coefficient, which allows it to

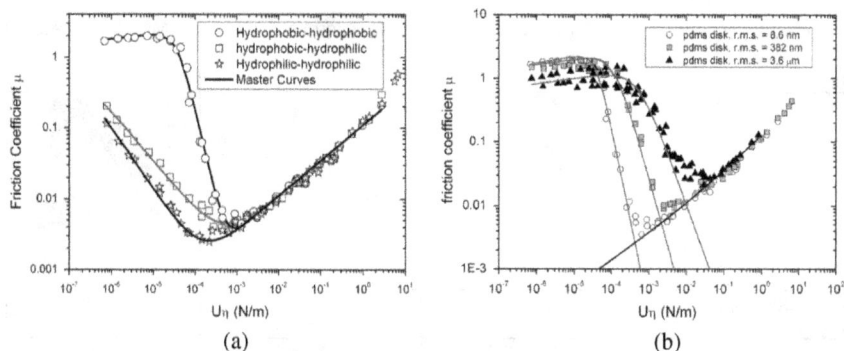

Fig. 3.9. Master Stribeck curve for a range of aqueous fluids between (a) hydrophilic, hydrophobic or mixed-PDMS ball-disk tribopairs and (b) hydrophobic PDMS ball-disk tribopair with varying disk surface roughness. Reprinted from Ref. 35 with permission from Elsevier.

develop a high viscosity under high pressure. In comparison, water is commonly regarded as a poor lubricant because of its low viscosity and its low pressure-viscosity coefficient, which means that its viscosity doesn't depend significantly on pressure. However, this is far from the case in soft tribological contacts because the contact pressure deforms the surface rather than induce a high pressure in the fluid so that the pressure-viscosity coefficient is not a factor on the fluids lubrication performance. Hence, in the elastohydrodynamic region, the same friction-speed (Stribeck) curve can be obtained for water if its viscosity is increased to that of oil through the addition of Newtonian thickening agents such as sugar or glycerol.[44]

There have been several studies attempting to understand and probe the response of emulsions in simulated oral contacts.[39,45,46] These have given valuable insights into the response of emulsions, particularly those studies that have utilised bio-surfaces such as a pigs tongue. For example, Dresselhuis *et al.*[39] concluded that their studies show a relationship between friction force measurements, in-mouth coalescence and fat perception (discussed below). De Hoog *et al.*[47] demonstrated that the sliding friction between a pigs tongue and oesophagus under load and speed for a model emulsion follow similar behaviour to that observed for glass-rubber surfaces, and discuss how the papillae on the tongue surfaces are compliant enough to deform under load. Due to the roughness of the tongue surfaces (see Fig. 3.10) under conditions that they explored, it is considered that the measurements were reflective of the boundary regime. Limited measurements are made on a range of food emulsions, but largely similar friction

Surface	Filiform papillae	Fungiform papillae	Microfabricated PDMS
Architecture	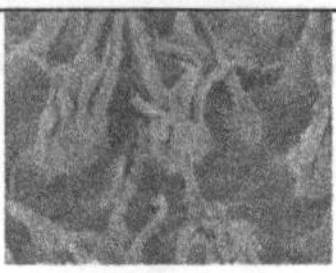		
Density, mm^{-2}	6 - 12	2 - 3	0 - 40
Diameter, μm	100 – 450	200 – 600	0 - 250
Height, μm	20 – 300	> 100	0 - 250

Fig. 3.10. Characteristics of filiform and fungiform papillae of the human tongue and microfabricated-PDMS consisting of hemispherical pillars.[48] Images reprinted from Ref. 48 with permission from Elsevier.

forces are observed for most of these against load for 10–40% oil stabilised with whey protein. However, the exception was cream containing 35% fat, which displayed a constant friction force with load.

Dresselhuis et al.[49] developed a novel tribological setup that allows sliding friction to be measured whilst viewing the contact with a confocal scanning laser microscope (Fig. 3.8). They used a pig's tongue as a model for the human tongue; piglets are used since they have closer degree of keratinisation to human tongue. However, long-term storage methods altered the structure and properties of the tongue surface, which is a major problem when using such 'real' surfaces. Friction measurements performed using the piglet tongues were also limited because they were susceptible to wear; this was because they were operated in the boundary regime where there was contact between the tongue surface and glass slide. Dresselhuis et al.[49] demonstrate that both the treated and dried tongue surface are intrinsically hydrophobic, although the presence of mucous fluids renders it hydrophilic. While the contact angle in air for water is similar to native hydrophobic surfaces, the 3-phase contact angle of water droplet in a bulk sunflower oil indicated that the PDMS is more hydrophobic under such circumstances. It is also observed that droplet coalescence occurs between the papillae, due to local destabilisation of the emulsion under shear. Shearing with smooth PDMS did not produce the coalescence that is observed for the very rough tongue surface; this may be expected purely due to a change in the roughness and not because PDMS is a poor mimic of the tongue surface. While the tongue surface operated in the boundary and mixed regime, smooth PDMS is typically operating in the hydrodynamic regime, where different behaviour is expected.

In order to more closely mimic the tongue surface, Ranc *et al.*[48] fabricated surface structures from PDMS that more closely resembled the roughness of the tongue surface (shown in Fig. 3.10), and examined the tribological properties using a reciprocating-motion sliding tribometer. The fabricated PDMS surface consisted of hemispherical pillars of varying diameter (100 μm and 250 μm), height (50–250 μm) and surface density. However, only a limited range of speeds and loads is examined and only a limited range of food emulsions. Under a single-speed condition, Bellamy *et al.*[50] compared iso-viscous emulsions at two different oil levels (22% and 33% MCT oil) and surfactant (non ionic sucrose ester and sodium casienate protein), and various levels of dextran as a thickener. This fat level was chosen because lower fat levels are not perceived by textural cues other than viscosity. However, the only difference for their samples (at the single speed tested) is for the 33% oil emulsion stabilised using sucrose ester, which led to a lower friction coefficient than the other emulsions; the reason for this is not established although the friction is higher than for the oil phase alone.

While the utilisation of real tissue surfaces and visualisation are extremely valuable, caution should be exercised because of the limited shear and load conditions currently available in these devices. Limited speed conditions and single-point measurements have the potential to lead to misinterpretations. For example, a decrease in friction coefficient from a base-line response could arise from several sources, including: (a) improved boundary lubrication; (b) an *increase* in viscosity if it is in the mixed regime; or (c) a *decrease* in viscosity if it is in the elasto-hydrodynamic regime. Nonetheless, these approaches provide a valuable insight into the behaviour of food emulsions under potentially relevant conditions to the actual process in-mouth, particularly when combined with an understanding on their behaviour using more traditional tribological approaches. They also emphasise that the major regimes in oral processing are likely to be the boundary and mixed regime, and that shear between asperities (papillae) and papillae deformation are also likely to be influential on the behaviour of food emulsions.

The utilisation of well-defined surfaces is essential to understand how complex microstructured fluids behave in tribological contacts. The surfaces in traditional tribometers can be easily manipulated to control roughness and surface chemistry, including wetting characteristics, so that a fundamental insight into how complex fluids behave can be obtained with precision. It is, however, important to also consider real tongue surfaces, with the view to either mimicking such surfaces and/or validating findings under well-controlled conditions that may more closely resemble those present in the mouth.

3.3.2. *Emulsions in soft tribological contacts*

In soft contacts, Malone *et al.*[51] reported the lubricating properties of complex emulsions comprising thickener (guar), surfactant and sunflower oil in between a hydrophilic steel ball and hydrophobic elastomer surface (see Fig. 3.11). However, different levels of guar are used to match the viscosity of the emulsions at a shear rate of $50\,s^{-1}$; this shear rate is chosen because it is considered to be the shear rate at which the sensory property of 'thickness' for viscous fluids is perceived. They found that the Stribeck curve varied with oil content; at low oil content, it didn't display any appreciable difference to a guar gum solution with no oil, while at 55% oil content the Stribeck curve follows that of the oil phase. At 15–30% oil, it was less clear and tended to be in between these two extremes of behaviour. As shown in Fig. 3.11, the sensory data showed that fattiness perception correlated well with oil content, but panellists found it hard to distinguish the 15–30% oil samples; it is suggested that this matched expectations from tribology measurements. While the data at 55% oil is similar to the data for oil, it is not precisely the same and sits above the oil data in the hydrodynamic regime. Therefore the viscosity of the sample in the contact is likely to be higher than the oil phase, which suggests a w/o emulsion may be present in the contact. This study highlighted that tribology may be an important attribute in the mouthfeel of emulsions. However, it is difficult to ascertain the mechanism for the observed tribological measurements. Further insight is needed to determine how different components lead to different Stribeck curves to enable the rational design of food emulsions. For example, there is no information on the rheology of the samples and so no way of knowing whether the observed friction data arises from the bulk rheology of the samples or localised phase separation (plating out) of oil in the contact zone. While the emulsions were viscosity matched at a single shear rate, these fluids are non-Newtonian and shear thinning; tribological contact is expected to be governed by significantly higher shear rates ($>1000\,s^{-1}$).

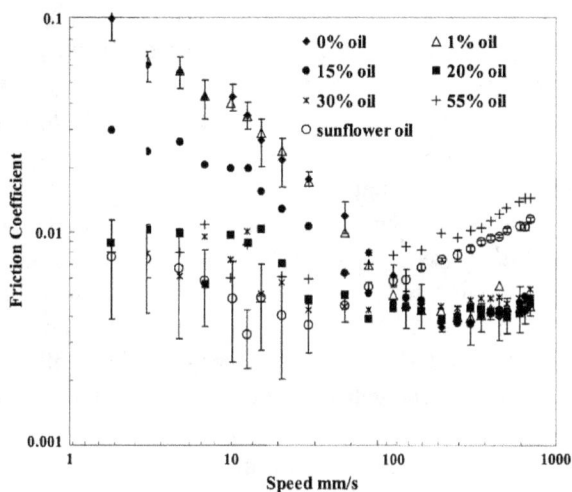

(a)

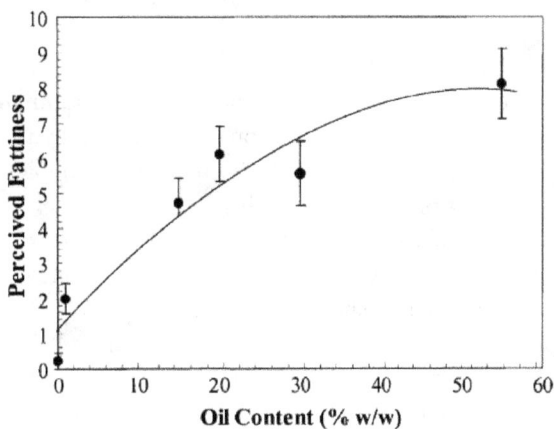

(b)

Fig. 3.11. (a) Friction coefficient measurements in an MTM between a steel ball and elastomer disk for a range of food o/w emulsions that differed in oil and thickener content so that they were isoviscous at a shear rate of $50\,\text{s}^{-1}$. (b) Sensory panel data on fattiness perception for the same emulsions. Reprinted from Ref. 51 with permission from Elsevier.

3.3.2.1. *Viscosity ratio*

Since foods contain a large range of ingredients with different rheologies, it is necessary to identify how these different viscosity phases affect the Stribeck curve. This was studied for the first time by de Vicente *et al.*[44] by increasing the viscosity of the water phase by adding glycerol or glucose syrup; these 'thickeners' were chosen because they are Newtonian and hence have a constant viscosity even at very high shear rates present within tribological contacts. The viscosity ratio (p) for o/w emulsions is given by $p = \eta_{oil}/\eta_{water}$, where η_{oil} is the viscosity of the dispersed oil phase and η_{water} is the continuous water phase. It is interesting to note that while the viscosity ratio is considered an important parameter for emulsion lubrication studies in engineering applications, it has not been specifically examined. The study by de Vicente et al.[44] used a hydrophilic steel ball rolling/sliding (slide-to-roll ratio of 50%) against a hydrophobic elastomer. A so-called master curve was obtained for a series of Newtonian aqueous fluids ranging in viscosity from about 1 mPas to 3 Pas, and sunflower oil. The sunflower oil sat on the same master curve and fell onto both the mixed and hydrodynamic regimes; this demonstrates that aqueous fluids can display the same lubrication behaviour as oil when matched in viscosity provided there are no significant surface active agents or boundary lubricants present (e.g. surfactants, fatty acids, etc.). This is an important realisation in soft-contact tribology. To avoid the influence of surface-active agents, no surfactant was added to the system and the emulsions were homogenised to obtain droplets of order $3\,\mu$m; this is of similar size to the gap at the minimum in friction coefficient corresponding to the junction between the hydrodynamic and mixed lubrication regimes. In this study, a mixed tribopair was used to avoid issues regarding preferential wetting.

For o/w emulsions without thickener, de Vicente et al.[44] found that the Stribeck curve resembled that of the oil phase for concentrations of 5% and above, but was closer to the water phase at 0.5 and 1% oil. Since both surfaces are preferentially wetted by either phase (DE < 0 for the elastomer surface, but DE > 0 for the steel surface), it is suggested that the oil being more viscous is able to displace the water phase from the contact. It is also noted that sunflower oil gave the same Stribeck curve when both surfaces were rendered hydrophilic. When the viscosity ratio is varied from $p = 47$ to $p = 0.09$ at a constant oil phase volume of 20%, as shown in Fig. 3.12, it is discovered that the Stribeck curves are mapped onto the master curve by multiplying the speed by an effective viscosity. These effective viscosities match either the oil phase or water phase. At high viscosity ratio, $p \geq 5.8$, the sunflower oil enters the contact and determines film formation and thus the friction coefficient, while at low viscosity ratios where $p \leq 1.3$, the aqueous phase determines the measured friction coefficient.

Recent studies have examined the influence of viscosity ratio further using emulsions with surfactant and surfaces with varying wetting

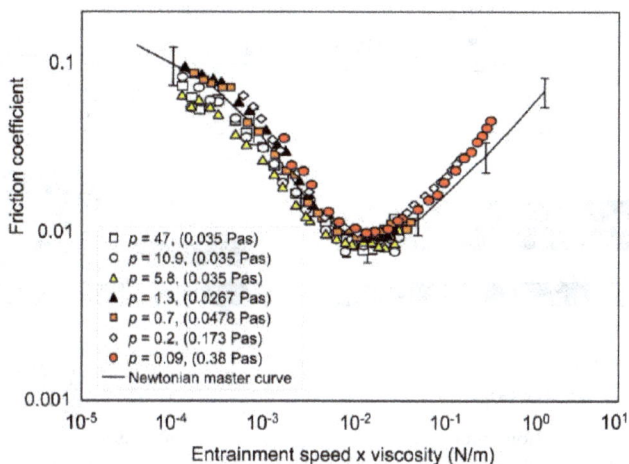

Fig. 3.12. Influence of viscosity ratio for surfactant-free o/w emulsions between a hydrophilic-steel ball and hydrophobic elastomer. The viscosity used to shift the data onto the master curve is shown and is equal to the thickened-water phase for $p \leq 1.3$ and to the oil phase for $p \geq 5.8$. Reprinted from Ref. 44 with permission from ASME.

properties. This is performed by using the same elastomer on both surfaces, i.e. a PDMS ball and PDMS plate that are natively hydrophobic but can be rendered hydrophilic using plasma treatment.[35] These studies are showing that the friction coefficient for emulsions without surfactant is determined by the phase that preferentially wets a symmetric tribopair, regardless of viscosity ratio. That is, the oil is preferentially pooled in the gap between two hydrophobic surfaces while the aqueous phase is preferentially pooled between hydrophilic surfaces; this recent finding demonstrates that the DE and work of adhesion approaches perfectly apply. However, for mixed tri-bopairs (hydrophobic and hydrophilic), the viscosity ratio largely controls the phase in the contact in a similar way to that observed previously,[44] as shown in Fig. 3.13.

The influence of viscosity ratio between phases in mixed–tribopairs can be understood in the context of general droplet coalescence studies. On approach of an oil droplet within an aqueous phase of high viscosity, it is more difficult to thin the aqueous film between the oil droplet and the substrate (and/or other oil droplets), thus reducing the potential for the oil droplet to wet the surface and thus coalesce in the contact zone. At low aqueous-phase viscosity, the film between droplet and surface/droplet will be easily squeezed out and hence show greater potential for coalescence

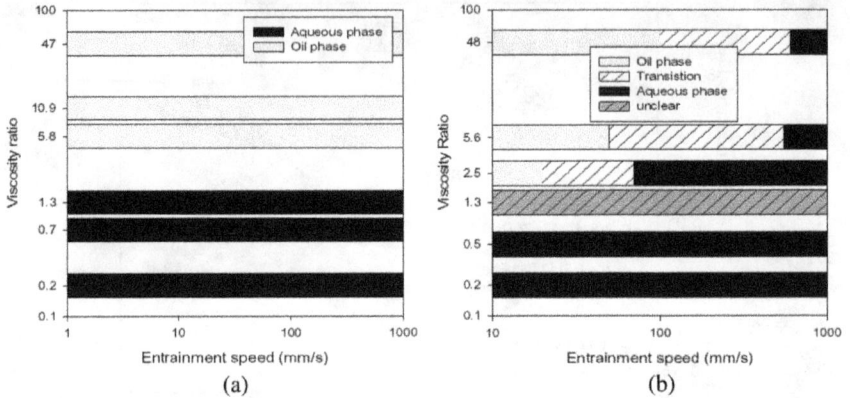

Fig. 3.13. Lubrication phase domain maps at different viscosity ratios for 20% *o/w* emulsions in soft-tribological contact: (a) smooth *hydrophilic*-PDMS disk and *hydrophobic*-PMDS ball without surfactant (b) smooth *hydrophobic*-PDMS disk and ball with surfactant (Note that for emulsions without surfactant, the domain map is dominated by oil across all viscosity ratios and speeds tested.)

or surface wetting in the contact. However, the wetting characteristics of the surfaces are also critical, and surfactants can be used to control and modulate this process.

3.3.2.2. *Non-ionic-surfactant-stabilised food emulsions*

Surfactants are used throughout foods to assist emulsion formation (by lowering the interfacial tension between the oil-water phases) and stabilisation (by providing a barrier to coalescence of the phases). In a tribological context, surfactants also provide the potential for formation of a self-assembled layer on the tribological substrates that alters their wetting and tribological characteristics. As previously highlighted, surfactants can render hydrophobic surfaces more hydrophilic, thereby enhancing fluid entrainment of aqueous fluids,[28] or vice-versa; the surfactant concentration and their ability to alter the wetting properties of the oil and aqueous phase has been shown to be crucially important.

For hydrophobic PDMS tribopairs, non-ionic surfactants enable the aqueous phase to wet hydrophobic substrates, but the viscosity ratio also has a major influence. At low entrainment speeds (mixed regime), the 'most' viscous phase (oil if $p > 1.3$, aqueous phase if $p < 1.3$) determines the friction in the contact as shown in figure 13. However, in the high-speed (thick-film) regime, the flow is typically governed by viscosity of the bulk emulsion or continuous phase, which is considered to arise from the surfactant-coated droplets depleting away from the shearing surfaces due to hydrodynamic forces. This process is apparent for droplets that are smaller than the film thickness between substrates.

Non-ionic surfactants have also been used to formulate and measure the macro-and nano-tribological properties of microemulsions[28]; microemulsions are thermodynamically stable emulsions ($\sim$10 nm diameter) that consist of micelles containing oil. It is apparent that the oil phase has very little effect on their lubrication properties, which are considered to arise purely from the adsorption of the surfactant onto the hydrophobic surfaces. This may also highlight the fact that large droplets are needed in order to obtain plating out of the oil phase, while very small droplets are more stable and likely to pass through or around the contact.

3.3.2.3. *Protein-stabilised food emulsions*

Many food emulsions (e.g. mayonnaise in Fig. 3.1) are stabilised by proteins, which are typically linear chains of polypeptides folded into globular structures that are surface active. Dresselhuis *et al.*[46,49,52] investigated the adhesion, wetting and spreading of protein-stabilised emulsions at surfaces varying in hydrophobicity as well as on porcine-tongue surfaces. Increased retention of fat/oil at solid surfaces was associated with a reduction of sliding friction between surfaces. Emulsion droplets stabilised by low amounts of protein are more prone to adhere and spread on tongue (and model hydrophobic surfaces) than more stable emulsion droplets stabilised with high amounts of protein.[39] This occurs due to increased hydrophobic attraction at short distances between the protein-poor emulsion droplets and hydrophobic surfaces due to decreased electrostatic and steric repulsion.[52] For example, using a flow cell it was shown that there is increased adherence and spreading of protein-stabilised emulsion droplets on a hydrophobic surface when electrostatic repulsion is decreased by increasing salt concentration.[52] Protein-rich emulsions are also less likely to break down at surfaces because of steric repulsion from a thick protein layer coating on the emulsion droplet.

Saliva is anticipated to play a key role in emulsion behaviour during oral processing, and has been observed to aid in the breakup of oil into droplets during mastication *in vivo*[53] and aggregate stabilised-oil droplets *in vitro*.[54] Figure 3.14 shows that saliva can increase the adhesion of whey-protein-stabilised oil droplets to the tongue surface, although upon washing

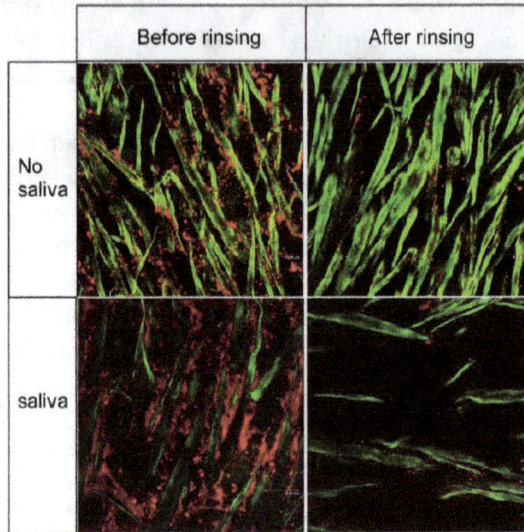

Fig. 3.14.   CSLM image of a pig's tongue in contact with a whey-protein-stabilised 10% o/w emulsion with and without saliva before and after rinsing with water. The red colour indicates the oil. Image width is 500 μm. Reprinted from Ref. 49 with permission from Elsevier.

with water most of the oil is removed. It is also observed that saliva can interact with adhered droplets to remove them from the surface.[52]

Dresselhuis *et al.*[39] evaluated the sensory properties of protein- and starch-stabilised emulsions, to test the influence of emulsion stability (against coalescence) *in vivo*, which also considers the influence of saliva. In an oral-processing context, coalescence may arise from the combined effects of shear and interfacial interactions from the oral surfaces, air and saliva. They developed a series of emulsions with various levels of protein emulsifier, as well as emulsions stabilised using an octenyl succinic anhydride (OSA) starch and solid fat. The novel aspect of OSA starch is that while it coats and stabilises oil droplets against coalescence in storage, it readily becomes unstable in the mouth because it is broken down by amylase in saliva. They found that emulsions with a higher sensitivity towards coalescence were perceived as most fatty and creamy, which confirmed their expectations from emulsion tribology and adhesion studies. Other sensorial attributes such as aroma release and mouthfeel were also affected by the emulsion stability. This research highlights how emulsion stability can be used to alter in-mouth perception; however, it is a challenge for food manufacturers who generally seek to have stable emulsions during manufacture, storage and transport. It also highlights that studying the behaviour of emulsions in tribological contacts and at surfaces leads to valuable insights that can be used for rational design of food emulsions to behave in a desired way in the mouth.

3.3.3. *Considering rheology*

Both fluid and soft foods are rheologically complex.[1] Food emulsions usually include at least one hydrocolloid (typically a polysaccharide) to increase the viscosity of the water phase and/or to give it solid-like properties below a yield stress; these serve to increase the time required for oil droplets to float (referred to as creaming) during storage and transport. The viscosity of the water phase, and the emulsion overall, is designed to have a high viscosity at low shear rates but the viscosity invariably decreases with increases in shear rate or shear stress, i.e. they are usually shear thinning. The rheology of foods and polymeric hydrocolloid solutions are usually characterised up to shear rates of $100 - 1000\,\mathrm{s}^{-1}$. However, in tribological contacts, fluids are subjected to shear in very narrow gaps where the lubricant experiences shear rates that far exceed those in standard rheological tests.

The influence of complex rheological parameters is yet to be studied in the emulsion-lubrication literature, but the influence of shear thinning on the lubrication of aqueous polymer solutions is briefly highlighted here. As the concentration of polymers is increased, it is found that the friction coefficient decreases substantially in the mixed-lubrication regime between a steel ball and elastomer disk in a mini-traction machine.[55,56] Using a standard rheometer, the high shear rates present in the tribological contact can be achieved using a narrow-gap, parallel-plate configuration.[57] By presenting the friction coefficient Stribeck curve as a function of $\eta_\infty U$, where η_∞ is evaluated using the viscosity at a shear rates of $10^5\,\mathrm{s}^{-1}$, it is found that most of the decrease in friction coefficient with increasing concentration is accounted for by the high shear viscosity alone. The result for aqueous solutions of xanthan gum (a polysaccharide that is commonly used to thicken food emulsions) is shown in Fig. 3.15, with the same findings also found for the polyethylene oxide solutions used by de Vicente *et al.*[55,56] The Stribeck curves at each polymer concentration overlay, but there is a slight lateral shift from the master curve that is ascribed to polymer adsorption onto the hydrophobic surface.[58] These results serve to highlight the importance of

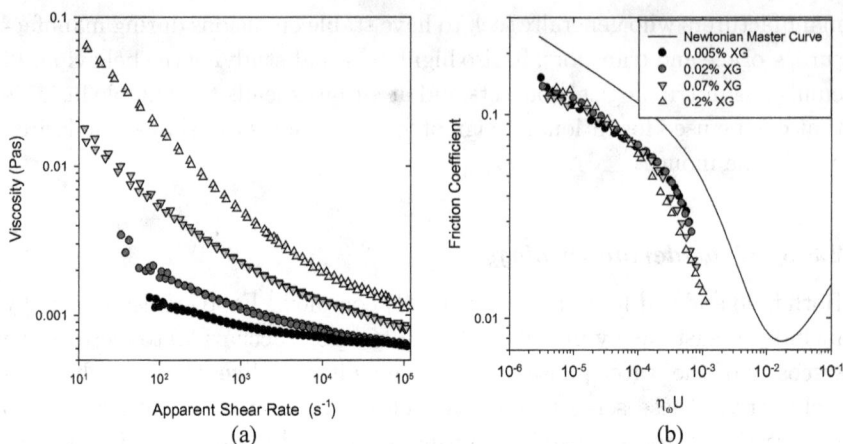

Fig. 3.15. Demonstration of the importance of rheological characterisation at the high shear rates present within tribological contacts. (a) Shear viscosity of xanthan gum polymer solutions up to shear rates of 10^5 s^{-1} using narrow-gap, parallel-plate rheometry. (b) Friction coefficient data that originally showed a decrease in friction with increasing concentration when plotted against entrainment speed[55,56] are replotted by multiplying the speed by the viscosity at 10^5 s^{-1}. The lateral shift in the polymer Stribeck curves is ascribed to polymer adsorption onto the hydrophobic substrate.

characterising the rheology at the shear rates estimated to be present in tribological contacts.

A strategy commonly used to replace fats in foods is to use water-thickening agents such as polysaccharides. However, mixed success with this strategy is potentially a result of the shear-thinning nature of these hydrocolloid solutions. While the 0.2% xanthan gum solution in Fig. 3.15 exhibits a viscosity similar to that of sunflower oil at shear rates around $10 \, \text{s}^{-1}$, it has a viscosity close to that of water at extremely high shear rates. Hence, it is anticipated that for a thickened water solution to have similar lubrication behaviour to oil, they would need to be viscosity matched at high shear rates.

3.4. Outlook for Emulsion and Food Emulsion Lubrication

The study of emulsions and food-emulsion lubrication is still largely in its infancy and a rich area for future study. Although there is some general understanding and knowhow, a predictive capability based on emulsion properties and surface characteristics is still in development. There appears to be consistency between the observations made in hard-EHD lubrication and those in soft-tribological contacts. However, while in engineering applications the focus is on the hard-EHD lubrication regime, the focus in foods research is towards oral tribology, which is principally in the boundary-mixed regime and features highly deformable, 'rough' surfaces. Foods are also highly complex in terms of microstructure, rheology and composition[1] and more research is still needed to both characterise and understand the lubrication properties of complex food systems under well-defined conditions and at biosubstrates. It is also necessary to understand the influence of saliva on food-emulsion lubrication. A major challenge in oral tribology is to ascertain how best to capture the complexity of oral substrates within *in vitro* measurement capabilities and to determine how tribological observations translate to in-mouth behaviour and/or to a sensory response. It will also be interesting to see whether understanding from the study of food emulsions may lead to new designs for emulsions in engineering applications.

Acknowledgements

I sincerely thank Jeroen Bongaerts and Juan de Vicente for their contribution and efforts towards the study of emulsion lubrication and soft-contact tribology, and Georgina Davies for her contribution to research on thin film rheology, which was performed while they were at Unilever's Corporate Research laboratory in the UK. I sincerely thank Prof Hugh Spikes at Imperial College London for many helpful past discussions, advice and collaborations. I also thank G. Cassin for introducing me to tribology, and the management team at Unilever for supporting research into this field including I. T. Norton, I. Appelqvist, and A.-M. Williamson. The tribology data in Fig. 3.13 was measured and obtained in collaboration with Jeroen Bongaerts (manuscript in preparation). The rheology data in Fig. 3.15 was measured by Georgina Davies.

References

1. J. R. Stokes and W. J. Frith, *Soft Matter* 4, 1133 (2008).
2. M. Ratoi-Salagean, H. Spikes and R. Hoogendoorn, *Proceedings of the Institution of Mechanical Engineers Part J-Journal of Engineering Tribology* 211, 195 (1997).
3. A. Azushima and S. Inagaki, *Tribology Transactions* 52, 427 (2009).
4. S. R. Schmid and J. Zhou, *J Tribol-T ASME* 123, 283 (2001).
5. S. R. Schmid and W. R. D. Wilson, *Lubrication Engineering* 52, 168 (1996).
6. A. Z. Szeri, *Wear* 200, 353 (1996).

7. S. W. Lo, K. C. Huang and M. C. Zhou, *Tribol Trans* 52, 66 (2009).
8. L. R. Ma *et al.*, *Colloids and Surfaces a — Physicochemical and Engineering Aspects* 340, 70 (2009).
9. J. C. J. Whetzel and S. Rodman, *Iron and Steel Engineer* 36, 123 (1959).
10. S. H. Wang, A. Alsharif, K. R. Rajagopal and A. Z. Szeri, *J Tribol-T ASME* 115, 515 (1993).
11. S. H. Wang, A. Z. Szeri and K. R. Rajagopal, *J Tribol-T ASME* 115, 523 (1993).
12. T. Nakahara, *J Jap Soc Tribologists* 40, 644 (1995).
13. T. Nakahara, T. Makino and K. Kyogoku, *J Tribol-T ASME* 110, 348 (1988).
14. D. C. Barker, G. J. Johnston, H. A. Spikes and T. F. Bunemann, *Tribol Trans* 36, 565 (1993).
15. D. Zhu, G. Biresaw, S. J. Clark and T. J. Kasun, *J Tribol-T ASME* 116, 310 (1994).
16. H. X. Yang, S. R. Schmid, R. A. Reich and T. J. Kasun, *J Tribol-T ASME* 128, 619 (2006).
17. H. X. Yang, S. R. Schmid, T. J. Kasun and R. A. Reich, *Tribol Trans* 47, 123 (2004).
18. Y. Shitara, *J Jap Soc Tribologists* 42, 540 (1997).
19. A. Cambiella *et al.*, *Tribol Lett* 22, 53 (2006).
20. A. Kumar, S. R. Schmid and W. R. D. Wilson, *Wear* 206, 130 (1997).
21. W. R. D. Wilson, Y. Sakaguchi and S. R. Schmid, *Tribol Trans* 37, 543 (1994).
22. Y. Kimura and K. Okada, *Tribol Trans* 32, 524 (1989).
23. A. K. Tieu and P. B. Kosasih, *Tribol Lett* 25, 23 (2007).
24. S. W. Lo, T. C. Yang, Y. A. Cian and K. C. Huang, *J Tribol* 132, 1 (2010).
25. M. Ratoisalagean, H. A. Spikes and H. L. Rieffe, *Tribol Trans* 40, 569 (1997).
26. K. Persson, P. Falk and J. Andersson, in *2nd European Conference on Tribology*, Pisa, Italy, (2009).
27. S. F. Chisholm, *Proceedings of the Institution of Mechanical Engineers* 179, 56 (1964–65).
28. M. Graca, J. H. H. Bongaerts, J. R. Stokes and S. Granick, *J Coll Interf Sci* 315, 662 (2007).
29. H. N. Patrick, G. G. Warr, S. Manne and I. A. Aksay, *Langmuir* 13, 4349 (1997).
30. L. M. Grant and W. A. Ducker, *J Phys Chem B* 101, 5337 (1997).
31. W. A. Ducker and L. M. Grant, *J Phys Chem* 100, 11507 (1996).
32. J. B. Hutchings and P. J. Lillford, *J Texture Studies* 19, 103 (1988).
33. J. L. Kokini, J. B. Kadane and E. L. Cussler, *J Texture Studies* 8, 195 (1977).
34. G. Luengo, M. Tsuchiya, M. Heuberger and J. Israelachvili, *J Food Science* 62, 767 (1997).
35. J. H. H. Bongaerts, K. Fourtouni and J. R. Stokes, *Tribol Int* 40, 1531 (2007).
36. S. Lee and N. D. Spencer, *Tribol Int* 38, 922 (2005).
37. J. de Vicente, J. R. Stokes and H. A. Spikes, *Tribol Lett* 20, 273 (2005).
38. G. Cassin, E. Heinrich and H. A. Spikes, *Tribol Lett* 11, 95 (2001).
39. D. M. Dresselhuis, E. H. A. de Hoog, M. A. C. Stuart and G. A. van Aken, *Food Hydrocolloids* 22, 323 (2008).

40. C. Myant and H. A. Spikes, *Proceedings of the ASME/STLE International Joint Tribiology Conference — 2009*, 59 (2010).
41. C. Myant, M. Fowell, H. A. Spikes and J. R. Stokes, *Tribol Trans* 53, 684 (2010).
42. C. Myant, T. Reddyhoff and H. A. Spikes, *Tribol Int* 43, 1960 (2010).
43. J. H. H. Bongaerts, J. P. R. Day, C. Marriott, P. D. A. Pudney and A. M. Williamson, *J Appl Phys* 104 (2008).
44. J. de Vicente, H. A. Spikes and J. R. Stokes, *J Tribol-T ASME* 128, 795 (2006).
45. T. van Vliet, G. A. van Aken, H. H. J. de Jongh and R. J. Hamer, *Adv Coll Interf Sci* 150, 27 (2009).
46. D. M. Dresselhuis *et al.*, *Food Biophysics* 2, 158 (2007).
47. E. H. A. de Hoog, J. F. Prinz, L. Huntjens, D. M. Dresselhuis and G. A. van Aken, *J Food Science* 71, E337 (2006).
48. H. Ranc *et al.*, *Colloids and Surfaces a — Physicochemical and Engineering Aspects* 276, 155 (2006).
49. D. M. Dresselhuis, M. A. C. Stuart, G. A. van Aken, R. G. Schipper and E. H. A. de Hoog, *J Coll Interf Sci* 321, 21 (2008).
50. M. Bellamy, N. Godinot, S. Mischler, N. Martin and C. Hartmann, *Int J Food Science and Technol* 44, 1939 (2009).
51. M. E. Malone, I. A. M. Appelqvist and I. T. Norton, *Food Hydrocolloids* 17, 763 (2003).
52. D. M. Dresselhuis, G. A. van Aken, E. H. A. de Hoog and M. A. C. Stuart, *Soft Matter* 4, 1079 (2008).
53. S. Adams, S. Singleton, R. Juskaitis and T. Wilson, *Food Hydrocolloids* 21, 986 (2007).
54. E. Silletti, M. H. Vingerhoeds, W. Norde and G. A. Van Aken, *Food Hydrocolloids* 21, 596 (2007).
55. J. De Vicente, J. R. Stokes and H. A. Spikes, *Food Hydrocolloids* 20, 483 (2006).
56. J. De Vicente, J. R. Stokes and H. A. Spikes, *Tribol Int* 38, 515 (2005).
57. G. A. Davies and J. R. Stokes, *J Non-Newtonian Fluid Mechanics* 148, 73 (2008).
58. J. R. Stokes, L. Macakova, A. Chojnicka-Paszun, C. G. de Kruif and H. H. J. de Jongh, *Langmuir* 27, 3474 (2011).

Chapter 4

Aqueous Lubrication in Cosmetics

Gustavo S. Luengo*, Anthony Galliano[†] and Claude Dubief*

L'Oréal Research and Innovation

Aulnay sous Bois, [†] Saint-Ouen, France

4.1. Introduction. The Importance of Aqueous Lubrication in Cosmetic Science

Cosmetic Science starts at the early stages of human development through the need to protect and the need to communicate. Cosmetic products are products intended to be applied to the human body to meet these needs and demands, including cleansing, beautifying, protecting and improving the condition of hair and skin. Cosmetics can be classified in various ways, although they are generally presented by function, e.g. skin care, make up, hair care, or fragrance.[1]

Water plays an important role in many of these products. Its role as a vehicle to deliver a particular activity or properties is one aspect, but it can be also considered as an active ingredient itself, when it is intended to improve skin hydration for example.

Lubrication, and in general tribological issues are involved in many cosmetic applications. Creams, for example, are expected to improve the texture and feel of skin; hair-styling lacquers to improve the adhesion between hair fibers to set a specific shape; nail varnishes to decrease wear in nails, etc. A soft feel and an even surface are often prime issues.

Aqueous lubrication phenomena are of particular significance in everyday shampoos and conditioners utilized for hair care. These highly evolved products are formulated to optimize hair appearance (shine, cleanliness), improve hair manageability and enhance hair feel. Sensory perception is therefore an essential property to be strived for.

When dealing with biological surfaces (hair and skin), the importance of the substrate cannot be underestimated; and understanding its importance is essential.

According to tradition, the purpose of hair is to protect what virtually all societies have considered to be the most precious and most noble part of the person: the head. Hair has always been one of the most powerful symbols of individual and group identity. Throughout many cultures, hair is seen as representing self-control, self-confidence, self-esteem, sexuality, distinction, freedom, and other qualities. In many ancient societies, men let their hair grow abundantly as a sign of power or election. In traditional societies, hair belongs to those ritual signs denoting status within a group: in modern societies that are more individualistic, hair has become an expression of the personality, providing a number of clues as to the tastes, profession, character, and opinions of the wearer. Everybody is different and choosing hairstyles that match each person will depend on these differences.

Hairstyle and manageability depends on numerous parameters: the geometry of the fibers (length, diameter, ellipticity, degree of curl), the topography of the hair coverage (covered head areas, density, angle of erection of the hair on the scalp), and fiber-to-fiber interactions, in other words, to a great extent on surface properties. The combination of all these parameters results in extremely diverse cosmetic properties. We can mention for example the clear tribological differences between dry hair and wet hair. Combing wet hair is difficult and that is why most conditioners are used after rinsing, when hair disentangling and manageability effects can be exacerbated. By a judicious selection of molecules, it is possible to modify fiber-to-fiber interactions and then to improve hairstyle.

This chapter will be structured in three parts: First, a general view of the known structure and properties of the hair surface and a description of the current understanding of the mechanisms of lubrication of common ingredients (surfactants and polymers), with a closer look at the molecular and physicochemical mechanisms concerning the adsorption of lubricating agents on keratin surfaces. Then, various methods to evaluate the effects on friction properties of a specific conditioning ingredient will be described, followed by an extensive review. Finally, we shall discuss some of the main technologies currently used to formulate hair-care products.

4.2. The Cosmetic Substrate

The main substrates of interest in cosmetic science are hair and the outermost layer of skin called the *stratum corneum*. Both are composed of dead cells that are mainly filled with specific types of protein called keratins. This protein is characterized by a high content of cystine, an amino acid that has the capacity to crosslink the protein by an intramolecular disulfide linkage. The structure and assembly of keratin proteins (and other components, such as lipids) in the cells and the cohesion of the cells give hair and the *stratum corneum* their specific properties, such as mechanical and tribological among others. Although the lubrication of skin is a very important issue in cosmetics, most formulations are based on an oil medium, which we shall not discuss here. Aqueous lubrication is naturally most related to hair cosmetics and that is why we have focused the rest of the chapter on hair applications.

4.2.1. *Hair*

Hair consists of more and less cylindrical fibers, roughly 50–100 μm in diameter. The overall number of hairs covering a normal human head lies between 120 000–150 000, which leads to a considerable surface area (typically $\sim 6\,m^2$ for $\sim 20\,cm$ long hair). This is a large surface to cleanse when shampooing, for example. Changes in the surface will therefore play a dominant role in the notions of perception and quality. Consequently, the knowledge and assessment of these properties are essential.

Hair is a good example of a bio-composite material, and it consists of three distinct morphological components: the outer protective layers known as the cuticle, the major structural components or the cortex, and a central, porous, irregular component called the medulla, which is not always present. Figure 4.1 shows a typical Transmission Electron Microscopy micrograph of a hair's thin microtomed section where cuticle and cortex can be easily observed.

4.2.2. *The cortex*

The cortex is the main part of the hair. It is formed of elongated cortical cells ($\sim 100\,\mu$m in length and ~ 1–$6\,\mu$m in diameter) aligned along the axis of the fiber and filled with partially crystallized keratins.

Hair keratin has the ability to crystallize in the form of an α-helical structure, basic element of a 4-strand rope arranged in a coiled-coil configuration.[2,3] These proteins have a lower cystine content (low sulfur proteins) and their chains are stabilized by many inter- and intramolecular interactions: electrostatic, van der Waals, hydrogen and covalent (disulfide bonds). These units form hexagonal crystals called microfibrils ($\sim 7.5\,nm$ in diameter) or intermediate filaments (IF). Moreover, these IF are embedded in an amorphous protein (matrix), constituting

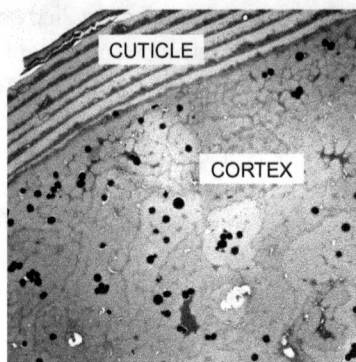

Fig. 4.1. Transmission electron micrograph of a thin transversal cut through a hair's shaft. The two main structures (cortex and cuticle) are clearly seen. The black spots correspond to the melanin granules. Notice the superimposed structure of the cuticle scales.

a semi-crystalline sub-structure of the cells called a macrofibril ($\sim$ 100–400 nm in diameter).[5]

These imbricate, crystalline, strongly anisotropic structures make the hair a true bio-nanocomposite, giving it unique mechanical properties: high rupture stress and elongation, high elastic modulus, rapid and total recovery even in some cases at high deformation.[5]

4.2.3. *The cuticle*

4.2.3.1. *The structure*

The cuticle forms the outer surface of hair. It is made of overlapping scales (approximately 50 μm long and 0.5 μm thick) protecting the cortex and oriented from the root to the tip of the fiber to produce a series of scales edges on the outer surface of each hair (Fig. 4.2). Figure 4.2 shows the aspect of the hair's surface using either Scanning Electron Microscopy (SEM) or Atomic Force Microscopy (AFM) techniques. With AFM accurate profiles of the scales are easily determined as is shown in the same Fig. 4.2. A normal hair has about 6–8 overlapping cells in close contact. In addition and thanks to earlier electron microscopy studies, we know that the cuticle includes three layers; the innermost endocuticle, the exocuticle and the outermost epicuticle. The endocuticle is believed to be the weakest component of the cuticle structure, where surface wear starts to propagate;

The final structure of the outermost layer of the cuticle is of special interest: the surface properties of hair depend on the physico-chemical properties of this layer.

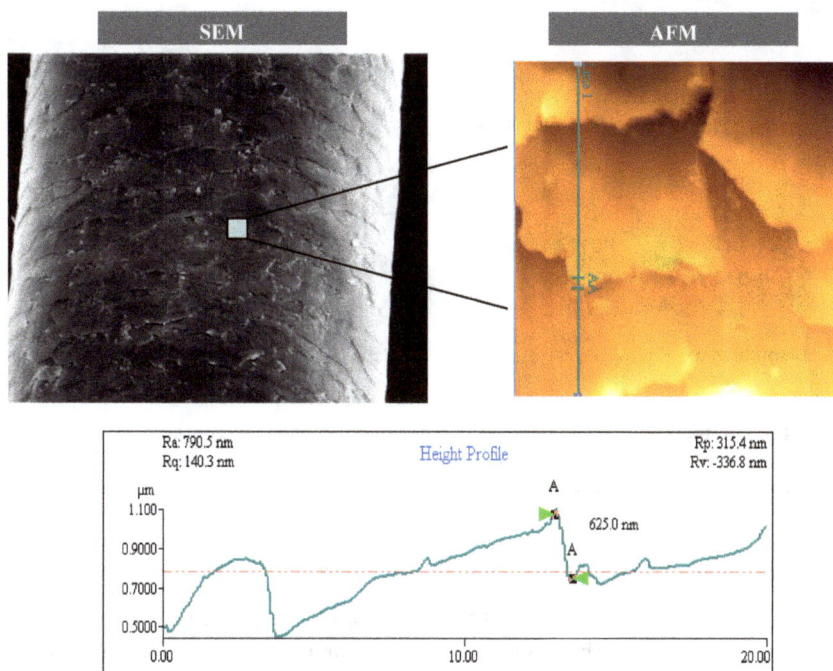

Fig. 4.2. Scanning electron micrograph (top left) of the hair surface. The scales are clearly seen. The fine topography is better observed on the profile (bottom), which shows the path indicated on the AFM image (vertical line).

The most recent findings have shown that the epicuticle membrane of hair fibers contains highly crosslinked protein ($\sim 75\%$) and fatty acids ($25\sim\%$). Among these adsorbed fatty acids, 18-methyleicosanoic acid (18-MEA) is the most abundant (50% w/w)[6] and it appears to be covalently grafted onto the outer surface (making the outer β-layer) via covalent thioester linkages to the protein.[7-10] It is considered to play an important role in the physico-chemical and tribological properties of hair.

Based on these observations, a model has emerged to explain the fine molecular structure of the hair surface.[11,12] Several studies, some of which have been initiated in our laboratories, have been performed to evaluate the accuracy of this representation.[13,14] As illustrated in Fig. 4.3, there are different ways for the 18-MEA layer to be grafted on the proteins of the epicuticule, either in a disordered (A) or ordered (B) pattern. In any case this layer seems to be more mobile than previously expected and does not fully cover the surface. In both situations, the presence of this lipid is associated with the excellent tribological properties of natural hair under ambient conditions.

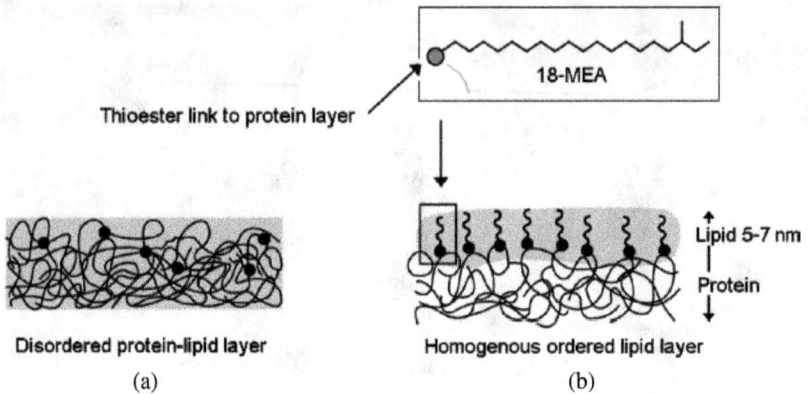

Fig. 4.3. A representation of the 18-MEA structure showing the ante — iso conformation (B, top). This lipid is likely to be found either forming a layer (B, bottom) or intimately associated with the protein (A).

4.2.3.2. *Function and importance of the cuticle*

Cuticle controls the diffusion of materials into the fiber, the bending stiffness, and because of the overlapping scales at the outer surface, facilitates the parallel alignment of the fibers on the scalp.[15] Its chemical reactivity depends on the physico-chemical characteristics of the surface, and especially on its wetting properties. Naturally, the lipidic layer of 18-MEA is responsible for the hydrophobic character of unaltered, natural hair.

The cuticle is the hair component most exposed to various grooming, environmental and chemical stresses during the life of a hair fiber. The cuticle cells at the outermost surface of the hair fiber are frequently subjected to harsh stresses, such as mechanical abrasion, UV sunlight exposure, bleaching, perming and blow drying. In fact we can distinguish chemical stresses, which alter the composition of the constituents of cells, and surface interactions and mechanical stresses, which break, or lift cuticle edges. The effect of these repetitive and most often additive stresses is cumulative damage to the cuticle, which is gradually disrupted, chipped, fragmented, and worn away.[5,16−20]

When this layer is modified, or disappears, especially as a result of oxidative treatments (bleaching, UV, etc.), keratin cystine linkages become oxidized into cysteic acid, which produces a more hydrophilic surface.[21−24] Table 4.1 shows the evolution of the contact angle with water in hair having undergone an H_2O_2 oxidation ("Lightly" or "medium bleached hair") or lye relaxer treatments ("Straightened hair") with rising intensity.

Table 4.1.   Average values of contact angle between water and hair with different "sensitization" levels measured by tensiometry.[25]

Natural Hair	Lightly Bleached Hair	Medium Bleached Hair	Straightened Hair
71°	65°	55°	42°

The reactivity of the surface therefore varies with hair "sensitization" ("sensitized hair" means hair which has undergone a permanent-waving, dyeing or bleaching treatment)[26] and determines, in part, the deposit of substances, such as cationic surfactants or polymers, on the hair surface,[27–30] or, for example, deposits of ceramide contained in a no-rinse product.[31]

Damage to the cuticle also increases as a function of time, and therefore it increases with increasing distance from the scalp. Sometimes the cuticle may be totally absent at the distal end, leading to the formation of split ends.

Damage that affects hair surface integrity increases diffuse reflection, decreases specula reflection, and impairs hair shine.[32–34]

4.3.   The Effect of water on hair structure

Water is involved at different stages during the application of a cosmetic product:

- Before application, to humidify hair. This step helps to prepare the product's "dilution" and promotes an even application over the entire head. After the product is dispensed, mechanical action helps to spread it and work it into a lather.
- During rinsing, which helps to rid hair fibre of excess product.

It is therefore worth noting that the performance of a product will not only depend on its reactivity with hair but also on the "effectiveness" of its application protocol on hair

Hair can absorb a large amount of water[35,36] and reacts very rapidly to changes in relative humidity, so much so that it was used in the manufacture of hygrometers for a long time.[37] Introducing water molecules into hair structure causes significant changes in the non-covalent bonds, hydrogen bonds and salt bonds, modifying mechanical properties, especially traction[38] (see Table 4.2).

Table 4.2. Average Young's Modulus values of simple extension of hair with different levels of "sensitization" and relative humidity (20%, 80% RH) and under wet conditions.

	Young's Modulus in Extension (MPa)			
	Natural Hair	Lightly Bleached Hair	Medium Bleached Hair	Highly Bleached Hair
20% RH	3658	3942	4189	4357
80% RH	2682	2811	2967	2933
Wet	1548	1482	1225	1133

On the whole, water acts like a plasticizer on hair. This evidence has also been observed at the microscopic level by means of nanoindentation.[39] Growing relative humidity causes the indentation Young's surface modulus to drop, especially in sensitized hair. Similarly, AFM measurements have shown that, in the presence of water, the cuticle, just like the cortex, swells by about 13%.[40]

However, water does not only affect hair "volume"; it also changes its surface properties and therefore its tribology.

4.4. Cosmetic Tribology. Lubrication Mechanism

4.4.1. *Fiber — fiber interaction*

The cuticle controls fiber-to-fiber interactions, which play an important role in fiber-assembly behavior, such as bending and compression of fiber bundles. When a fiber assembly is subjected to deformation strains, single fibers slip past one another at their contact points, thus minimizing the extent of deformation of individual fibers. The efficiency of this process depends on fiber-to-fiber friction, which arises as a result of adhesion force at the contact points.[41]

These interfiber forces play a significant role in combing ease,[42] hair breakage during combing,[43] body, flyaway, manageability, and style retention of hair assemblies. They also play a role in the overall appearance of the whole head of hair. Excess sebum, for example, can stick together several hairs to form little swatches, which characterize greasy hair.

Recent studies using AFM have allowed the interaction between two fibers to be measured in both dry and in aqueous environments.[44,45] The results show the importance of in-contact (tribological) and out-of-contact (interaction) forces on the dynamics of two fibers (compared to that of a fiber against a comb, a hand, etc.) and the effects of a cosmetic ingredient.

This type of experiment also helps modeling efforts to mimic real-life hair movement,[46] expanding the view we have of the visual impact of our ingredients.

4.4.2. Hair tribology

One of the distinctive features of friction on hair is associated with its structure. Hair's surface consists of scales laid out much like roof tiles. This organization first results in *anisotropy*, which reflects on the measurement of friction coefficient, depending on the direction of measurement. The coefficient is higher when sliding takes place from the tip to the root ("against the scales"). Generally, we also observe a relatively characteristic stick-slip phenomenon, as shown in Fig. 4.4.[47,48]

Naturally, the level of friction depends on the condition of the hair surface. As we have seen before, the surface of "natural" hair consists of a lipid layer of 18-MEA that contributes to friction.[49] All action inhibiting or deteriorating this layer, or modifying the surface (deposit[50]), will change friction properties.

4.4.3. The effect of water on hair tribology: A bad lubricant

Scientific literature has shown that friction coefficient tends to rise when hair is wetted or ambient humidity is high.[51,52]

For a better understanding of the phenomena, we can compare two friction curves, obtained on a single hair in dry and wet conditions. The changes in amplitude and form of the friction curves are noted in Fig. 4.5.

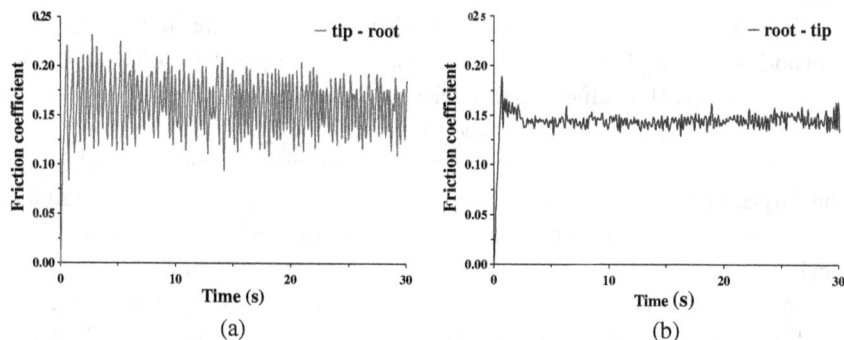

Fig. 4.4. Friction on hair: the value of friction coefficient depends on the direction of friction, (a) tip-to-root direction, (b) root-to-tip direction.

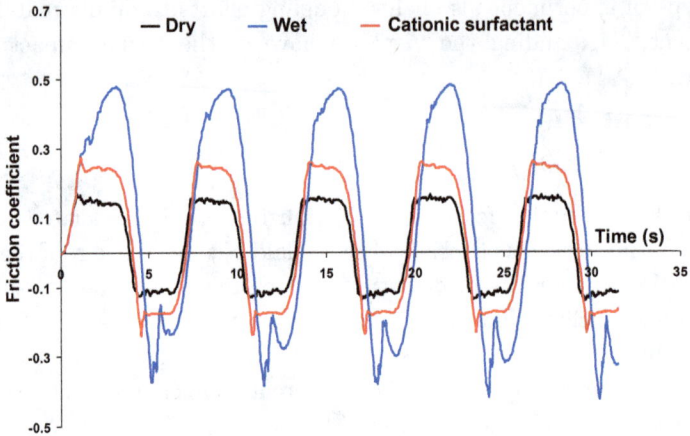

Fig. 4.5.   Measurement taken on a bleached hair with a normal load of 20 mN and a stainless steel cylindrical 1 mm ϕ friction block under dry (black curve) and wet (blue curve) conditions, and with an added cationic surfactant (red curve).

Under wet conditions, the value of friction coefficient is nearly multiplied by 3 and a very significant static peak appears. In addition, the "plateau" of friction is highly limited under wet conditions; the plasticization and adhesion effects are thus predominant in this state.

With the addition of a *cationic surfactant*, the friction coefficient drops drastically to the level of dry hair. The sliding plateau reappears and the presence of the static peak is attenuated. The lubrication effects on the interface are highly significant and the deviation with the level of dry friction may be associated with the role of plastic deformation of hair in contact.

The same type of behaviour can also be found using another type of method, for example with *swatch-on-swatch* sliding (see Sec. 4.5.2.2). Under wet conditions, the value of hair's friction force increases significantly, even more than when hair is "sensitized" (see Table 4.3).

At the same time, it is also interesting to observe the evolution of combing forces over the course of a blow-drying cycle (Fig. 4.6). As a reminder, the swatch is initially wet and an operator proceeds with manual blow-drying using a brush and a hair dryer. Starting with the first strokes of the brush, the force is relatively high, between 5 and 15 N, depending on the applied treatment. With time when hair is still wet, the curve goes through a maximum point when hair breakage is at its peak. Later, with increasing drying time, the combing force drops and becomes stabilized.

Table 4.3. Values of the friction force (N) obtained with swatch-on-swatch sliding under dry and wet conditions for natural and bleached hair with *Les*Na shampoo.

	Sliding Force (N)	
	Dry	Wet
Natural hair	0.43	0.95
Bleached hair	0.37	1.22

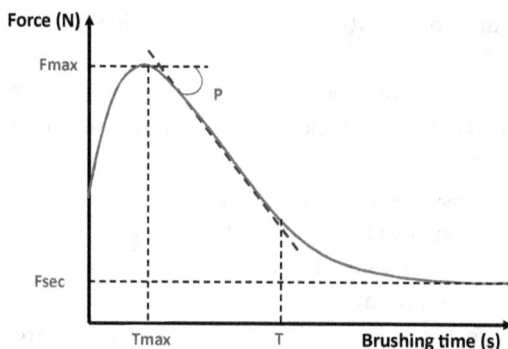

Fig. 4.6. Follow-up shape of the combing force during a blow-drying test.[53]

4.4.4. *Lubrication mechanisms of polymers and surfactants*

During bleaching, the oxidants used in this process to solubilize and eliminate melanin pigments have two side effects: they partially remove the fatty acids covalently linked to the outer surface of the hair cuticle (18-methyleicosanoic acid mainly present in the F-layer of the cuticle); they disrupt cystine linkages of the keratin and oxidize them to cysteic acid. As a result, in water, the ionization effect makes the fiber more negatively charged and more receptive to the charge-driven process for a positively charged adsorbing species.

This is why cationic surfactants and cationic polyelectrolytes are widely used. Together with silicones they are the main compounds currently used to provide lubrication to hair. (In the last part of this chapter a detailed description of the ingredients and products will be presented.)

In many hair-care products, polyelectrolytes are not used alone but in a mixture with surfactants. For example, in a shampoo a cationic polyelectrolyte will be generally present with oppositely charged surfactants, it is to say anionic surfactants.

In bulk water, polyelectrolytes interact with oppositely charged surfactant micelles to form, depending of the concentration of the charged compounds, soluble complexes, precipitates (liquid-solid phase separation) or coacervates (liquid-liquid phase separation). Coacervation is a phenomenon, in which a macromolecular aqueous solution separates into two immiscible liquid phases. The denser phase, which has a relatively high content of macromolecules, is called the coacervate and is in equilibrium with a relatively dilute macromolecular phase.[54] For polyelectrolyte-surfactant systems, coacervation yields a phase rich in both polyelectrolyte and surfactant.[55−57]

Many factors influence polyelectrolyte-micelle coacervation. Among these are polyelectrolyte properties, such as charge density, nature and position of the cationic charges, molecular weight, concentration, hydrophilicity and chain stiffness, presence of hydrophobic moieties; micelle properties, such as surface-charge density (nature of the anionic surfactant, presence of a co-surfactant), size, and concentration; and polymer-to-surfactant ratio, ionic strength, and temperature.

The surface deposition and phase behaviour of mixtures of anionic surfactants and cationic polyelectrolytes with different charge densities and hydrophobicities were recently investigated by A.V. Svensson *et al.*[58] The role of these two parameters on the delivery of emulsified silicone oil to hydrophobic silica was also discussed.[59]

By choosing the proper combinations of these different factors it is possible to obtain one-phase compositions at shampoo concentrations with phase separation occurring upon dilution during the shampooing process, leading to the deposition of conditioning agents — coacervates — on the hair.

The concept of forming and depositing complex coacervates onto hair is thought to be the key mechanism which provides cosmetic conditioning but little is known about the actual structure and mechanism at the bio-interface to explain this lubricating effect.

Because of their importance in the development of these products, a number of studies have been carried out using other techniques such as electron spectroscopy for chemical analysis (ESCA). Yet these techniques hardly give visual proof of the presence of adsorbed polymers (e.g. dye or fluorescence techniques).

4.4.4.1. *Nano-tribology approach*

There are several techniques of special interest for the understanding of the physics of cosmetic agents on surfaces. Some of them are specially suited for flat surfaces while others can be applied to complex topographic surfaces like hair. Like any other surface studies, it is by a clever combination of these techniques that we can make advances in elucidating the right chemical structures present in an aqueous environment, and thereby develop an efficient conditioning agent.

First, the surface forces apparatus (SFA)[60] is still the most quantitative method to measure surface forces. SFA allows the interaction between two symmetric mica surfaces to be assessed. The distance is measured by an interferometric method. Light reflects back and forth between the back-silvered mica surfaces generating fringes whose shape corresponds to that of the contact geometry, thus allowing the real contact surface area to be measured. The use of SFA provides a useful complementary tool for understanding the adsorption processes of shampoos and conditioners from water. The quartz crystal microbalance (QCM) is another technique of interest to understand the adsorption kinetics and wet adsorbed mass of the conditioning agents, as well as its mechanical properties. QCM is based on the changes in the resonant frequency of a quartz crystal, and is used to quantify the thickness and the amount of wet mass adsorbed onto a surface. Finally ellipsometry is a well-established technique that is used to measure the dry thickness of adsorbed films.

It is the availability of atomic force microscopy (AFM)[49,61] that has enabled much easier access to the study of physicochemical phenomena on the hair surface.

The instrument is based on the interaction of a sharp point (tip) attached to the end of a cantilever with the surface of interest. High-resolution topographical images are easily obtained under many environmental conditions and in particular in controlled humidity or water (an advantage over scanning electron microscopy (SEM), which is generally used under vacuum conditions).

Beyond imaging, AFM offers a wide range of possibilities to analyze mechanical (i.e. hardness) and physicochemical properties, based on adhesion (normal) or friction (lateral) measurements,[62,63] possibly combined with chemical modification of the hair ends (grafting hydrophilic or hydrophobic chemical agents).

The LFM or FFM mode[64] measures the friction forces exerted by the tip on the hair surface. In order to do this, scanning is performed perpendicularly to the axis of the cantilever. The friction induced by the lateral tip/sample contact leads to torsion at the end of the lever. At a nanoscopic scale, as in the case of contact between the AFM tip and the surface, the friction force is proportional to the normal load applied and the relation is linear. It is hence possible to determine the friction coefficient on a microscopic scale.[65,66]

The applications of AFM to cosmetics are relatively numerous, in general to study the topography, mechanics and tribology of hair. It is especially suited, for example, to determine the scale heights,[40,67,68] to visualize thin deposits of cosmetic treatments,[39,69] to study the influence of ethnicity, damage, or environmental conditions, or to measure interactions between hairs.[70–72]

4.4.4.2. Adsorption mechanisms

The mechanisms of lubrication with cationic surfactants are relatively well understood based on years of fundamental studies performed on flat mica surfaces, many of them using the SFA. The general conclusion is that a self-assembly process of the cationic heads on the negatively charged surfaces creates a "boundary lubricant" situation where the surfactant tails point away from the surface. The situation can be complicated, depending on the medium, if multilayer structures can be obtained (i.e. bilayers). Depending on the robustness of this layer the surfactant can act as a protector of the surface, thus diminishing friction.

Polymer adsorption becomes more complicated. A description of the parameters that control polymer precipitation, adsorption and structural rearrangement is beyond the scope of this chapter (some of this situations are presented in other chapters of this book). We shall nevertheless explain that an understanding of the structure and conformation at the surface is essential, in order to deduce properties/chemical structure relationships that eventually will help in the design of better conditioners.

One of the first challenges is the visualization of the conditioning agents, which form very thin films on the surface. An early application of AFM as a visualizing tool involved using mica as a model surface for keratin (both being negatively charged). Goddard and Schmitt[73] immersed a mica surface in a 0.1% polyquaternium-10 solution (Polymer JR400, a typical conditioning polymer) and observed complete coverage as reflected by a definite increase in the surface roughness. It is interesting to note that SEM images were unable to detect the presence of the polymer.

Since conditioning polymers for hair are frequently applied in anionic-surfactant-based shampoo, the authors further made the system more complex by incorporating small amounts of anionic surfactant.[74] They observed different types of texture as a function of the polymer used or the concentration of the solution. In general the adsorbed polymer layers were less homogeneous than when adsorbed alone, suggesting a complexation of the polymer with the anionic surfactant.

Many of these subtle variations are dependent on the right choice of the model surface. This is an issue of great importance, because any model surface may resemble hair in certain physico-chemical aspects (contact angle, surface potential, etc.) but not in all. It is also true that the flatness of the mica substrate allows us to concentrate on the adsorption dynamics of the material.

This complex interaction between cationic polymers and surfactants has been further and more recently investigated on real hair fibers.[77] The use of non-contact modes and phase images as well as the capability of repositioning and washing *in situ* helped the authors to attribute the net-like structures they observed to the polymer-surfactant complexes.[75,76]

The adsorption of these polymer-surfactant complexes changes the tribological properties of the hair fiber; for example, recent results show that shampoos and conditioners may affect adhesion at the nanoscale, and Blach *et al.*[78] have reported a clear decrease in the pull-off force on wet fibers.

Our laboratory has worked extensively on these kinds of problems. Using the SFA, the normal and friction forces between flat surfaces of mica carrying an adsorbed layer of Polyquaternium-10 were measured in the presence or absence of (a) a cationic surfactant and (b) a cationic and non-ionic surfactant together,[79] helping to understand the relation between the chain extension into the water medium and the friction forces observed. Other experiments with AFM have been focused on the possible bilayer structural organization of a quaternary ammonium surfactant on the hair fiber.[80]

Finally, the influence of more innovative structures can be illustrated by using copolymers with a tendency to adopt a brush-like structure (see Fig. 4.7). This is the case for the well-known poly(lysine)-poly(ethylene glycol) copolymer (PLL-g-PEG), which has been studied extensively in the past. Recent studies using hair as a substrate[81] have confirmed the critical role of the polymer structure, which that proved to improve not only the wet friction of hair but also the feel of the fibers.

| Surfactant_like | Random-gel like | Grafted PEG-like |

Fig. 4.7. Cartoon illustrating our understanding of the different polymer conformations present on the hair surface that are at the root of the lubricating effects observed with conditioning materials. Cationic surfactants are expected to form ordered thin films while common cationic polyelectrolytes adsorb in a more random gel-like structure.

4.5. Lubrication evaluation

Many instrumental tests are implemented, in order to assist the formulator in screening new molecules, and, more pragmatically, in developing claims based on product activity and performance.[82-84] These tests must be discriminating, while ensuring that measured properties are perceived by the consumers. They are designed based on three main themes:

- The "tactile" perception, which brings together all measurements relative to the surface and interface phenomena, (e.g. the feel), as well as tests focusing on "cosmetic properties" directly related to everyday life (brushing, combing, etc.),
- The visual perception, which includes measurements linked to the characterisation of appearance (colour, shine), form (volume, style) and movement (the "bounce" in hair, the "helmet effect", etc.),
- The integrity of fibre, which brings together tests concerning fibre quality (often tests of mechanical characterisation, traction, flexion, torsion tests, etc.)

Different physical, physicochemical and spectroscopic approaches may be used.[85] In this chapter, we focus on those related to tribology.

4.5.1. *From the material properties to the performance measurement*

Lubrication evaluation may be approached *directly* through the determination of intrinsic physical properties:

- The characterisation of the fibre changes generated by hair cosmetic treatments, pollution, or simply photodegradation caused by sun exposure[16,20,86−89]; or the deposits generated by various products (shampoo, conditioner or dye).

- The study of interactions among hairs: the behaviour of a head of hair may be associated with the combination of the mechanical properties of individual hairs as well as with interactions among hairs.[5,42]

or, *indirectly*, through tests more closely associated with consumer behavior. These tests will help evaluate the product's performance:

- Brushing, combing or detangling: the notions of the interface are more difficult to analyze because they involve contact between a brush or a comb and hair. Adding cosmetic products will directly change the tribological behaviour of these systems.
- Wear and tear: this phenomenon is often the result of cosmetic routine. Fibre damages following mechanical fatigue or friction, may lead to changes in the "feel" (sensation of "coarse" hair), or even breakage.

4.5.2. *Evaluation tests: direct and indirect measurements — from micro to macro*

In order to identify tribological hair tests, we can approach friction from a multi-scale perspective:

- from microscopic measurements, which typically explore at the scale of a single hair and help to directly describe material and deposit properties; In this case intrinsic properties are easily accessible using AFM (see nanotribology section)
- then at the scale of a hair swatch or head, where "indirect" tests are developed to help measure cosmetic performance ensuing from tribology measurements.

4.5.2.1. *Material characterization at the single-hair scale*

At the single-hair scale, measuring the friction coefficient is difficult. Different geometries can be used[90] and it is not so simple to relate "cosmetic" properties with a variable that is not intrinsic to the material (since it depends on the nature of the antagonist material: hair/hair, swatch/swatch, swatches friction, etc.).

Two test geometries can be used to measure tribological properties:

(1) The "capstan" method":

This system is shown in Fig. 4.8.

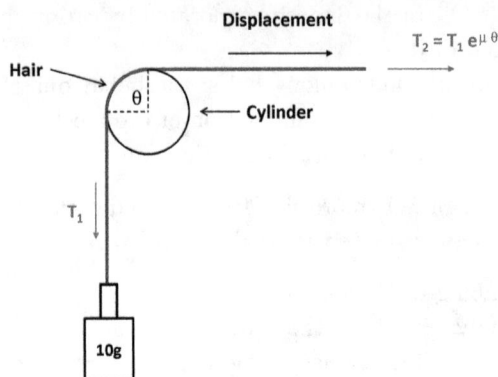

Fig. 4.8.   Friction system with capstan geometry.

A hair, to which a 10 gram weight is attached, is pulled at a low speed on a cylinder covered with paper. This surface helps to obtain a reproducible signal and reduce the contamination of the friction block (in the case of transfer, it is relatively simple to renew the surface). Measuring the necessary force to generate and maintain movement yields the friction coefficient of hair, given by the relation:

$$\mu = \frac{1}{\theta} \ln \left(\frac{T_2}{T_1} \right)$$

where T_2 denotes the registered friction force (resistance to advancement)
    T_1 is the tension applied to hair
    θ is the contact angle of hair on the cylinder ($\approx \pi/2$)
    The main disadvantage of this system lies in creep problems, which may appear during measurements under wet conditions. In addition, the minimal load in this type of configuration and the contact surface area between hair and cylinder remain relatively high. In order to avoid these drawbacks, another type of geometry may be employed:

(2) A "crossed cylinder" measurement method:

This system is compatible with that of AFM, where the measurements of tangential and normal forces are performed with a cantilever with known stiffness values. The principle and appearance of friction curves are shown in Fig. 4.9.

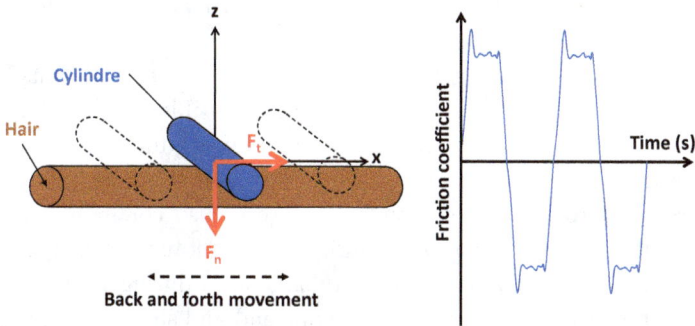

Fig. 4.9. "Crossed cylinder" measurement method. The hair and the friction block are in a crossed position (left). With each return cycle, the system records the tangential force and calculates the friction coefficient (right).

The friction coefficient is obtained with the classical frictional law of Amontons, where F_n is the applied normal force and F_t the measured tangential force:

$$\mu = \frac{F_t}{F_n}$$

The measurement process is carried out as follows:

- The slider comes in contact with the hair and the normal contact force (F_n) is obtained by compressing the cylinder on the hair.
- The hair moves by completing cycles of amplitude L at a frequency determined by the operator. During friction, the lateral deflection of the cantilever is recorded, in order to determine the friction force F_t. In order to maintain the normal force F_n constant during testing, a piezoelectric system adjusts, in real time, the vertical deflection of the cantilever with a feedback loop.

The normal force exerted on the hair, the friction speed, the number of cycles and the amplitude of movement will influence the friction process linked to the tribological properties of the tested substrate:

- At a "high" normal force (20 mN), it is possible to describe gross hair sensitization or the cosmetic performance of products by characterizing the [hair + deposit] system. At a "low" normal force (0.2 mN), "only" the deposit is characterized (smoothing/blocking). It can be empirically estimated that the range of applied load in this test is in the same order of magnitude as that applied by the finger during rubbing, i.e. from 0.4

to 4N (40 to 400 g), depending on the operator and the mode of strain (orthogonal or tangential to the surface, etc.).

- By increasing the number of cycles, hair loses matter and signs of wear and tear begin to take effect.[91] It is then possible, by correlating surface images, to demonstrate fibre-protection effects during the application of treatments. One test was specially developed to study this problem; it is referred to as flexabrasion.[92] This test, also commonly known as the "bending test", helps to quantify the resistance of a hair subject to repeated strain, similar to that taking place during brushing, which leads to breakage. Hair is fastened at one end and slides over a tightened stainless steel wire. A mass is attached to the other end and maintains a constant effort of traction over the fibre. Hair is driven with a rectilinear back-and-forth movement that results in breakage. Hair fatigue is characterized by the number of movements (back and forth) undergone before it breaks.[93] The greater the damage to the hair, the faster it breaks. Conversely, cosmetic treatments (lubricating) may increase the life cycle by reducing fibre friction on the cord, or by reinforcing the cortex.[94] Resistance is also higher with high humidity plasticizing hair.[95,96]

4.5.2.2. *From the hair swatch to the head: cosmetic performance*

Determining a product's performance may be relatively subjective. The ideal solution is to develop methods similar to consumer routines or to correlate the measured properties with sensory evaluations.

(1) Wet friction measurement:

The "swatch-on-swatch" sliding test is used to characterize the condition of hair surface. It consists of measuring the necessary force to slide a swatch of hair between two other swatches at a constant speed, as shown in Fig. 4.10.

This relatively simple and effective test proves pertinent in formulating different hair-care claims.[90] It is in fact correlated with the perception of a smooth feel to the touch after product application under wet conditions.

(2) "Combing" measurements:

There are several kinds of combing measurements, which are all intended to reproduce phenomena of everyday life. The goal is to demonstrate the performance of a product in reducing breakage induced by hair friction

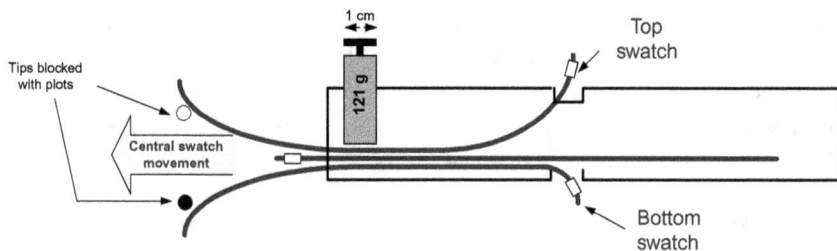

Fig. 4.10. Swatch-on-swatch sliding test. A mobile swatch is pulled in an even, rectilinear movement between two other immobile swatches positioned head-to-tail. The necessary force to pull the swatch is then measured with a force sensor attached to a driving arm where the swatch is positioned.

inside one or more combs.[43,97−99] All these tests begin with the swatch of hair being wetted:

(a) Garcia combing:

A swatch of hair, cleaned with a cosmetic treatment (shampoo, conditioner, etc.), is wetted, combed and then tangled according to a standardized protocol. The combing forces on the swatch are measured with an extensometer. The swatch is fastened, at the root end, to the force sensor and then engaged into a comb. The swatch passes through the comb from root to end at a constant speed. The force exerted on the swatch during traction is measured.[100,101] Changes can be made by adding a second comb,[102] increasing speed and performing combing cycles.[103] As a result, the "combing ease" corresponds to the sum of efforts over the course of these cycles.

(b) Detangling:

The measurements are taken with a detangling machine consisting of two vertically mobile combs that form a "jaw" opening and closing over a hair swatch. The swatch is fastened to the force sensor and the "jaw" moves vertically in the open position, closing over the hair with the teeth of the comb overlapping, before being lowered in the closed position to detangle the swatch. The detangling occurs in successive segments, starting at the bottom. Every time, the combs are raised by another grade level and lowered to the bottom of the swatch. After the last level, the combs pass through the entire length of the swatch. The force exerted by the combs to detangle the swatch is recorded at each level. When the force exerted on the swatch exceeds a certain threshold set by the experimenting investigator, the machine is blocked. This point of resistance is considered as a knot. In this case, the combs are opened and repositioned above the knot to resume their lowering movement (several times if necessary). The number of back-and-forth movements of the comb required to undo the knot is counted. The ease of detangling is obtained through the total forces recorded at each level and the number of times the comb is required to pass through the swatch to detangle the knots.

(c) Breakage during blow-drying:

This test helps evaluate changes in hair during blow-drying. The goal is to reproduce the thermo-mechanical strain consistent with a blow-drying routine by repeatedly brushing a swatch of hair that had been previously treated with a hair care product, while drying it with a hair dryer. Broken hair is recovered and weighed. The more weakened the hair, the more it will tend to break.[103] A product that reduces breakage can thus be considered as protective against blow-drying.

(3) The static electricity phenomenon:

This phenomenon is often present during hair brushing. Triboelectrical phenomena generate charges on the hair surface that create repulsive forces between hair fibres.[104,105] As a result, the hair is difficult to style. The phenomenon does not occur at high relative humidity.

Based on the methods described above, the following chapter specifically discusses evidence of lubricating effects provided by cosmetic products under wet conditions.

4.5.3. *From intrinsic to functional properties*

As described before, water plays a role in different stages of the treatment, in the application and distribution of the product over hair, and in rinsing. We can more closely examine these different phases by illustrating our observations with actual product examples, for example, regarding the performance of products under wet conditions.

The facts are therefore clear: friction rises significantly when hair is wet. In addition to washing hair (ridding fibre of impurities, sebum, pollution, etc.), cosmetic products must provide "treatment" benefits (helping to detangle, smooth hair, reduce breakage, etc.).

In order to better understand the different ranges of performance that products belong to, we have correlated different shampoo and conditioner technologies with different techniques.[106,107]

Table 4.4 shows the results obtained with wet sliding, blow-drying and wet combing methods for leading shampoo and conditioner technologies, and for two-product (shamp+cond) and three-product (shmp + cond.+ treatment) combinations.

Table 4.4. Average values of wet sliding forces, masses of broken hair during blow-drying and average combing forces (*Shp.*: shampoo — *Cond.*: conditioner).[108]

"Swatch-on-Swatch" Sliding Test Under Wet Condition (Force N)		Brushing (Weight g)		Wet Combing (Force N)	
Shp. 1 Sodium Laureth Sulfate (SLES)/Coco Betaine	1.15	*Shp.* Sodium Laureth Sulfate/Coco Betaine	19.5	*Shp.* Sodium Lauryl Ether Sulfate	1.37
Shp. SLES/Disodium Cocoamphodiacetate/Guar Hydroxypropyl Trimonium Chloride/Dimethicone	1.02	*Shp.* SLES/ Cocamidopropyl Betaine/ Polyquaternium 6/Amodimethicone	11.1	*Shp.* SLES/ Cocamidopropyl Betaine/ Polyquaternium 10	0.67
Shp. SLES/Coco Betaine/ Polyquaternium 10	0.77	*Cond.* Amodimethicone/ Hydroxyethylmonium methosulfate/ Behentrimonium chloride	7.9	*Shp.* SLES/Coco Betaine/Guar Hydroxypropyl Trimonium Chloride/Dimethicone	0.5
Shp. SLES/Cocobetaine/ Polyquaternium 10/Amodimethicone	0.69	*Binome =* Shampoo + conditioner	6.9	*Cond.* Behentrimonium Chloride/ Amodimethicon	0.22
Cond. Hydroxyethylmonium methosulfate/Cetrimonium Chloride/ Amodimethicone/ Behentrimonium Chloride	0.61	*Trinôme =* Shampoo + conditioner + lay on conditioning spray :	2		

(Continued)

Table 4.4. (*Continued*)

"Swatch-on-Swatch" Sliding Test Under Wet Condition (Force N)		Brushing (Weight g)	Wet Combing (Force N)
Cond. Quaternium 87	**0.55**	Dimethicone/ Amodime-thicone/ Polyquaternium 4/Behentrimo-nium Chloride	
Cond. Amodimethicone/ Cetrimonium Chloride/ Stearamidopropy-ldimethylamine	**0.52**		
Cond. Behentrimonium Chloride/ Amodimethicon	**0.37**		

The different shampoo and conditioner technologies are described in the next Section 4.6. Only the main active ingredients of the formulae are reported in Table 4.4.

On the whole, we can observe two measurement ranges where the sliding forces for conditioner technologies are on average twice as low as the shampoo technologies. The "strength" of treatment or the use of two or three product combinations generally decreases the friction forces.

These results confirm those already obtained in leading lubricant categories, such as cationic surfactants, polymers and silicones (PDMS), contained in shampoos and conditioners.[109,110]

Based on this fact, it is understandable that deposit quality and properties will determine product performance.

4.6. Hair Care Products: Ingredients and Formulation

The desire to have beautiful, healthy hair leads cosmetic scientists to develop and improve hair-care products. Two families of products are used by most people on a daily basis: shampoos and conditioners.

The first function of a shampoo is to clean the hair by removing excess soil and sebum. However, shampoo can leave the hair in a wet, tangled, and generally unmanageable state. Once the hair dries, it is often left in a dry, rough, lustreless, or frizzy condition, mainly due to removal of the hair's natural lipids. The hair can further be left with high level of static electricity upon drying, which can interfere with combing.

In this state hair fibers are most prone to be damaged by abrasion.

A variety of approaches has been developed to alleviate after-shampoo problems of the hair.

These range from post-shampoo application of hair conditioners, for example hair rinse, which is applied in a separate step following the shampoo, left on the hair for a length of time and then rinsed out. Other post-shampoo conditioning aids have been developed, such as mousses, gels, and lotions. These are also applied to the hair in a separate step but not rinsed out.

Another solution is to provide hair-conditioning benefits in shampoos.

In order to obtain the desirable properties, a wide variety of conditioning agents has been studied and optimized in specific compositions. The major classes of conditioning agents used in commercial products and their principles of formulation are surveyed in the following sections.

4.6.1. *Raw materials*

4.6.1.1. *Ingredients*: *Cationic surfactants*

Cationic surfactants, in the form of quaternary ammonium salts, are widely used as conditioning agents in commercial products. They are very effective on damaged hair, especially on bleached hair. There are many cationic surfactants. For conditioning purposes, the longer aliphatic chain compounds are more effective than those with shorter aliphatic chains. A more hydrophobic cationic surfactant leads to increased deposition on hair. This is true for mono aliphatic chain surfactants as well as for surfactants having two or three aliphatic chains.

Important examples of these compounds include:

- Quaternary ammonium salts with one or two aliphatic chains (Fig. 4.11)
- Quaternized or non-quaternized amino amides of fatty acids, such as stearoyl aminopropyl dimethylamine (Fig. 4.12)
- Fatty amines, such as stearyl dimethylamine
- Quaternized or non-quaternized ethoxylated fatty amines (Fig. 4.13)
- Esterquats, which comprise a class of quaternary salts having the general formula $R_4N^+X^-$, where the hydrophobic part of the radical R, which

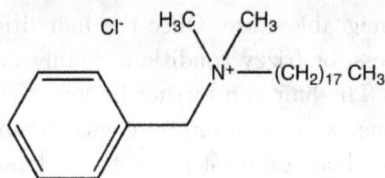

(a)

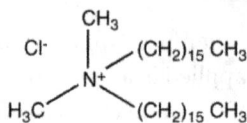

(b)

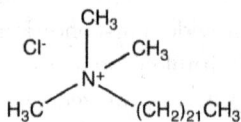

(c)

Fig. 4.11. Stearyl dimethyl benzylammonium chloride (a), dicetyl dimethyl ammonium chloride (b), behenyl trimethyl ammonium chloride (c).

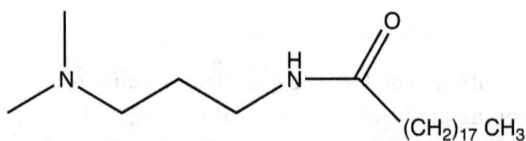

Fig. 4.12. Dimethylamine or stearamidopropyl dimethyl amine.

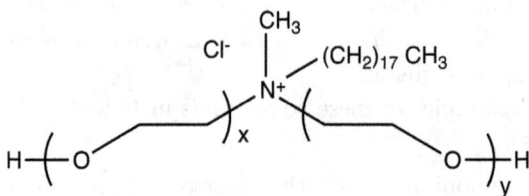

Fig. 4.13. PEG-2 stearmonium Chloride where x+y has an average value of 2 (INCI name).

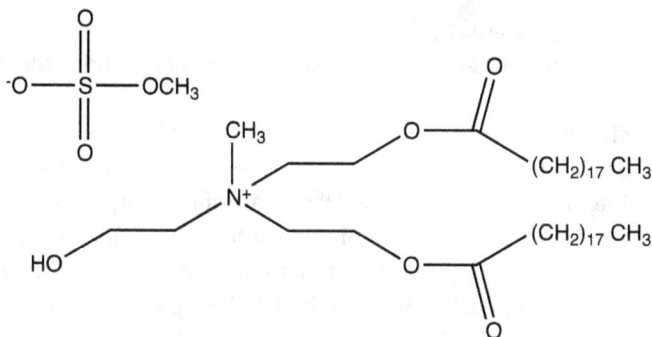

Fig. 4.14. Distearoylethyl hydroxyethyl methylammonium methosulfate.

contains more than four carbon atoms, is linked to the positively charged head group via an ester bond (c.f. Fig. 4.14).

Concern for the environment has led to the synthesis of these last compounds, which exhibit good conditioning properties and increased biodegradability and environmental safety. One such example is shown in Fig. 4.14.

Cationic surfactants are never used alone. The great majority of commercial products known as rinses are combinations with long-chain fatty alcohols, especially cetyl and stearyl alcohols. The relative amounts of cationic surfactant and fatty alcohols will determine the appearance and the rheology of the dispersions, but also their level of care. These mixtures have been found to form liquid-crystal and gel networks that determine the rheology and stability of the dispersion.[111–115] It was also found that the addition of cetyl alcohol to behentrimonium chloride composition resulted in significantly reduced surface friction.[52]

4.6.1.2. Lipophilic conditioners

Providing hair with suitable fatty materials to lubricate, protect and lend it softness and sheen has been prized in all civilizations. Traditionally olive oil, karite oil, ratanjot oil, shorea oil, monoi oil, jojoba oil, or avocado oil have been used by local populations.

Currently, to complement these natural oils, the most frequently lipophilic compounds used in commercial products are:

- fatty acids (oleic, stearic, behenic acids...),
- fatty alcohols (lauryl, myristyl, oleyl, cetyl, stearyl alcohols...),
- other natural triglycerides (almond, castor, peanut, corn oils...),

- natural waxes (beeswax, jojoba oil...),
- fatty esters, such as glycol stearates and synthetic short alcohol fatty esters,
- oxyethylenated or oxypropylenated waxes, alcohols, and fatty acids,
- natural or synthetic ceramides, which are fatty acylated sphinganine, sphingosine, or phytosphingosine.[116,117] Among them, N-oleoyl sphinganine[118] has been shown to restore the integrity of damaged hair cuticle and re-establish its protective function.[31] In particular, it maintains the cohesion of the cuticle when submitted to physical stresses such as brushing.[119]

As mentioned earlier, the great majority of hair conditioners consist basically of cationic dispersions of waxes, mainly fatty alcohols.

These products treat, beautify and repair hair, generally by depositing the conditioning agents on the surface of the hair. For a greater, more lasting effect, it is sometimes desirable to transport those agents into the deepest layers of the cuticle. This can be done employing formulations with a vector effect.

Oil miniemulsions, especially those containing vegetable oils, are representative of hair-care products with a vector effect.

Emulsions can be considered as mini emulsions when their overall particle size is below 500 nm in diameter.

The word "miniemulsions" covers all thermodynamically unstable emulsions with "small" droplets irrespective of the way they are produced.

High-pressure homogenization (HPH) is one mechanical method of obtaining miniemulsions. In this technique, a primary coarse emulsion is compressed up to 1500 bar and forced through a homogenizing valve. The flow of products is transformed into one or several spray streams that converge in a mixing chamber. The extreme shear rate in the mixing chamber creates very small droplet sizes.

The presence of cationic surfactants within the emulsifying system of the miniemulsion increases its vector capacity.

4.6.1.3. *Cationic polyelectrolytes*

Although cationic surfactants are very efficient at smoothing out the cuticle scales, imparting softness, facilitating disentangling, combing, and brushing and protecting damaged areas, their formulation has some limitations. Many of them are not compatible with anionic surfactants and lose their conditioning effects in the presence of these species, such as salts of alkylsulfates or alkylethersulfates — the surfactants most commonly used in commercial shampoos.

Cationic polyelectrolytes have made a breakthrough in conditioners and have given the way to the use of conditioning shampoos. In these products, the cationic polyelectrolyte is incorporated at low concentration ($<1\%$) in the composition, which usually contains 10 to 20% of anionic surfactant.

Examples of cationic polyelectrolytes used as conditioning agents in shampoos are:

- Polyquaternium 10 (INCI Name): Fig. 4.15
- Polyquaternium-67 (INCI Name): Fig. 4.16
- Guar hydroxypropyl trimonium chloride (INCI name):
  Is a hydroxypropylated guar gum modified with 2,3-epoxypropyl trimethylammonium chloride or with 3-chloro-2-hydroxypropyl trimethylammonium chloride to introduce cationic charges into the molecule.
- Polyquaternium-6 (INCI Name): Fig. 4.17
- Polyquaternium-16 (INCI Name): Fig. 4.18
- Polyquaternium-46 (INCI Name): Fig. 4.19

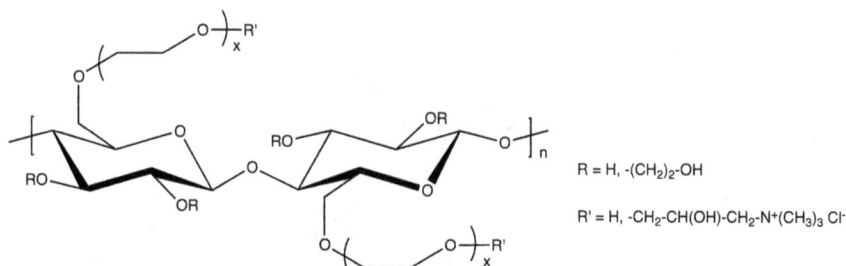

$R = H, -(CH_2)_2-OH$

$R' = H, -CH_2-CH(OH)-CH_2-N^+(CH_3)_3\ Cl^-$

Fig. 4.15. Hydroxyethylcellulose quaternized with 2,3-epoxypropyl trimethylammonium chloride.

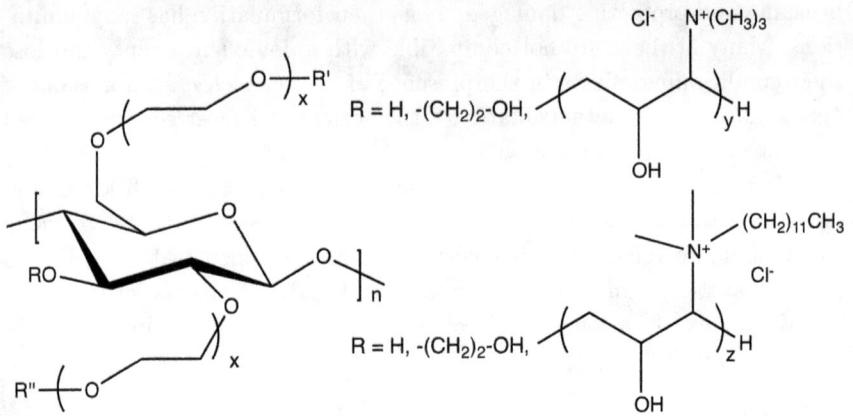

Fig. 4.16. Hydroxyethylcellulose quaternized with 2,3-epoxypropyltrimethylammonium chloride and 2,3-epoxypropyldimethyldodecylammonium chloride.

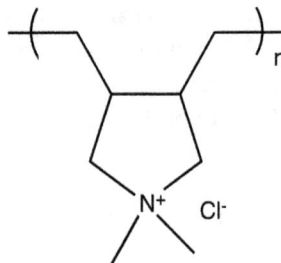

Fig. 4.17. Polydiallyl dimethyl ammonium chloride.

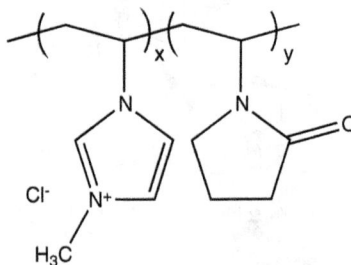

Fig. 4.18. Copolymer vinylpyrrolidone/methyl vinylimidazolium chloride.

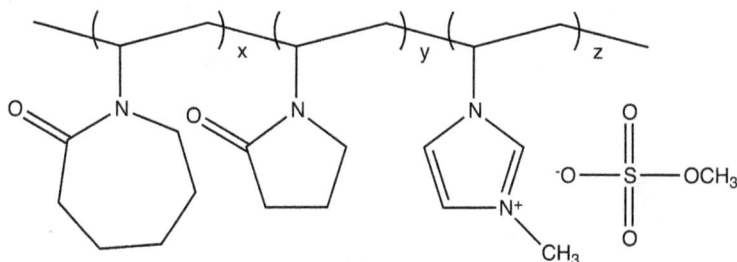

Fig. 4.19. Terpolymer vinylcaprolactame/vinylpyrrolidone/methyl vinylimidazolium.

4.6.1.4. *Silicones*

The use of silicones in hair cosmetics dates back to the 1950s. Silicones cover a wide range of architectures, including silicone homopolymers, silicone random copolymers and silicone-organic block copolymers

Polydimethylsiloxanes are the largest-volume silicone homopolymers produced to date. They exhibit the characteristic properties of the silicone family: chemically inert, low surface tension, low glass-transition temperature, weak cohesive forces, and water insolubility. This causes them to spread easily on most surfaces to give a quite uniform, smooth, hydrophobic deposit. Several silicone homopolymers find use in hair-care products:

1. Cyclic siloxanes Fig. 4.20(a): The most common cyclic siloxane used in cosmetics is decamethyl cyclopentasiloxane (where n = 5). It is mainly used in "rinses", combing aids and conditioners.
2. Linear dimethylsiloxane with trimethylsiloxy end group Fig. 4.20(b): A large variety of polymers belongs to this family, with viscosities ranging from 0.65 to 1 000 000 mm^2/s at 25°C. The silicones with lower viscosity (approximately 4 mm^2/s) are volatile.
   Silicones having higher viscosities appear like oils or gums and are insoluble in water but soluble in aliphatic hydrocarbons. Dimethicones with medium and high viscosities (50 000 to 1 000 000 mm^2/s and more) are extensively used as conditioning agents in shampoos.
3. Aminofunctionalized silicones

The most commonly used are copolymers of dimethylsiloxane and (di) amino alkyl methylsiloxane such as: Trimethyl silylamodimethicone (Fig. 4.21(a)), and Amodimethicone (Fig. 4.21(b)).

Amodimethicone differs from trimethyl silylamodimethicone by its reactive silanols groups. A wide range of polymers with various molecular weights and amine groups are available. The presence of amine groups in

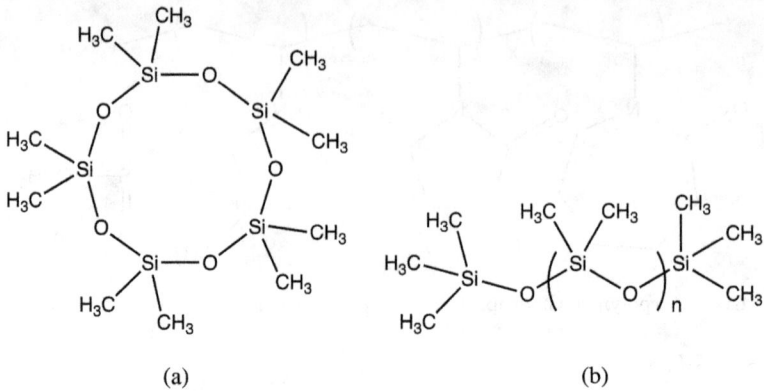

(a)   (b)

Fig. 4.20.   Decamethyl cyclopentasiloxane (a), Linear polydimethylsiloxane with trimethylsiloxy endgroups (b).

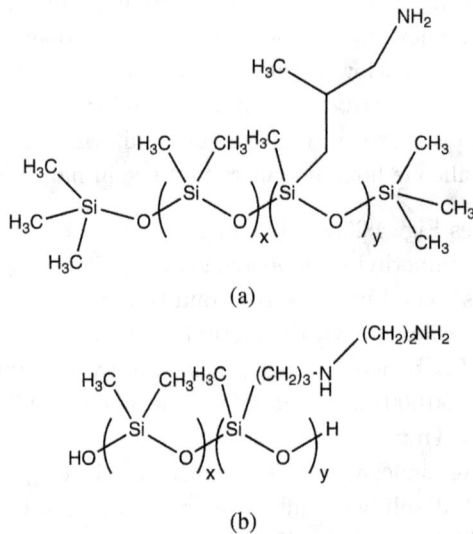

(a)

(b)

Fig. 4.21.   Trimethylsilylamodimethicone (a), amodimethicone (b).

silicone structures increases the hydrophilicity of the polymer and makes it easier to formulate. When the silicone derivative has a high content of amine groups, neutralizing with an organic or inorganic acid can make it soluble in aqueous medium. Furthermore, silicones with amine groups are cationic polymers with good affinity to hair. They are important components of conditioners and shampoos.

The exact mechanisms leading to the adsorption from a shampoo of dispersed silicone onto hair have not been totally described. It is shown that the deposition of silicone oil droplets on hair depends on many factors: the nature of the ionic charge of the droplet, the droplet size,[120,121] the presence of competing adsorbents, the presence of adsorption promoters, the hydrophilicity of the hair (virgin vs damaged hair).[122]

Hair fiber is amphoteric. In its virgin state, its isoelectric point is about pH 3.7. Below this pH, hair is positively charged and above it is negatively charged. Above this pH, aminofunctional silicone and cationic emulsions of silicone are generally found to adsorb to a greater extent than anionic or non-ionic emulsions of silicone. Experiments with hair and silicone emulsions show that the larger droplets deposit better than smaller droplets.

A series of quaternized hydroxyethylcellulose with trimethyl ammonium and dimethyldodecyl ammonium groups (Fig. 4.16) have been evaluated in shampoo formulations containing a dispersed silicone. It was found that increasing the charge density of the polymers above 1.45 meq/g resulted in a several-fold boost of their co-deposition ability.[123]

In the case of damaged hair, bleached hair or hair colored by a permanent dyeing process, all of which correspond to hydrophilic fibers, it is shown that polydimethylsiloxane deposition is reduced and does not provide the same hair-conditioning benefit as on virgin or lightly damaged hair. This is particularly perceptible for silicone-containing shampoos.

Several approaches have been used to increase the efficiency of conditioning shampoos on damaged hair. The best solutions involved cationic polyelectrolytes.

It has been found that specific grades of cationic guar gum increase silicone uptake when associated with silicone emulsions in hair shampoo formulation.[124]

In a very interesting paper[125] dealing with cationic polyelectrolyte affinity to hair during shampoo application, foaming and rinsing, new issues not really taken into account by physical chemists were pointed out. First, shampoo dilution is not an equilibrium phenomenon but a complex kinetic process involving water flow, mechanical shear, and foam removal. Second, during these different phases, depending on the composition of the formulae, coacervates, aggregates, and precipitates can be formed and adsorbed onto the hair surface. Third, coacervates, precipitates and other insoluble particles present in the shampoo, partition within the foam between the wall of the foam lamella and the lamellar liquid. Fourth, in this study it was found that the preferential partitioning or adsorption of coacervates onto the foam lamella depended on their surface activity while that of the insoluble particles depended on their size and their hydrophobicity. Finally it is suggested that during foam rinsing, flocculation of the various colloidal structures takes place, giving rise to the formation of heterogeneous aggregates that adsorb onto the hair surface.

Surprisingly it was found[128] that the presence or absence of insoluble components in a shampoo change the way in which the cationic polyelectrolyte interacts with the hair surface. This is the case for some cationic hydroxyethylcelluloses (Fig. 4.15). When the shampoo contains polydimethylsiloxane, the deposition of the polycation is smooth and uniform; in contrast, when the shampoo does not contain silicone, the deposition of the polyelectrolyte is not uniform and concentrates at certain specific spots onto the hair surface. In addition, the affinity of the polycation to hair is higher when the shampoo contains silicone.

4.6.2. *Conditioning agents and their formulation*

Hair and scalp accumulate a wide diversity of soil including sebum, dead epidermal cells, residues from hair dressings, hair sprays, as well as dust and grime from the air. Soiled hair lacks luster, becomes oily and unmanageable and develops an unpleasant odor.

The first purpose of a shampoo is to clean the hair and the scalp but in order to answer to the consumers' demands, shampoos are now designed to provide, in addition to soil removal, ease of wet and dry combing, manageability, body, control of static charge, luster, soft feel and so on.

The components used to formulate shampoos fall into three classes:

(1) The cleansing base,
(2) The conditioning agents,
(3) The additives that confer the viscosity, appearance, feel, perfume, and stability to the products.

Some optional ingredients, such as vitamins, vegetal extracts, antidandruff, moisturizing agents, antiparasitics, etc., can also be added for specific purposes.

4.6.2.1. *Cleansing base*

The surfactants most often used are the C_{12}-C_{14} alkylsulfates and alkylethersulfate (2-3 EO) due to their excellent foaming and cleansing properties. Commonly, these surfactants are combined with an amphoteric surfactant, such as C_{12}-C_{14} alkylbetaines or C_{12}-C_{14} alkylamidobetaines. Mixtures offer many advantages such as tighter, softer, more stable foam. They can also modulate the formation of coacervates. Sometimes non-ionic surfactants can be used for this last purpose.

Surfactants of the cleansing base represent 10–20% (in weight) of the formula.

4.6.2.2. *Conditioning agents*

Cationic polyelectrolytes, and silicones are the most widely used components in the shampoos.

Cationic polyelectrolytes are water soluble and when used at low concentration (0.1 to 1%), they are generally compatible with the other components of a shampoo.

The use of silicone is more complicated. In shampoos, the best conditioning effects are obtained when the silicones are insoluble during the rinsing step. A single way to obtain this result is to use silicones incompatible with the ingredient mix of the shampoo. But in these compositions, phase separation into layers occurs. It is of course highly desirable to formulate stable bi-phasic compositions where silicones are uniformly dispersed as discrete droplets throughout the products.

The problem has been solved by incorporating specific thickeners such as ethyleneglycol distearate and stearic monoethanolamide,[126] or xanthan gum.[127]

Silicone usually represents 0.1 to 3% of the total composition.

Currently, in order to optimize the conditioning performance of the shampoo, cationic polyelectrolytes and silicones are often used together.

4.6.2.3. *Other additives*

A shampoo should be viscous enough to stay in the hand before application, yet during application it must spread easily and disperse quickly over the head and the hair.

The thickeners for shampoos are chosen for their compatibility with the surfactants mix. Most widely used are:

- electrolytes, in particular sodium chloride and/or ammonium chloride and the salt of citric acid,

- fatty acid amides and more specifically the alkanolamides such as ethanolamides and isopropanolamides,
- polysaccharides, whether chemically modified or not,
- synthetic polymers such as cross-liked homopolymers of acrylic acid.

The final touch given to the shampoo is an appropriate perfume, pearlescent or opacified appearance, and coloration if desired.

The greatly used pearlescent and opacifier agents are ethyleneglycol stearate and ethyleneglycol distearate, the amides of C18 (or greater) saturated fatty acid, the cetyl, stearyl, behenic alcohols and their derivatives, and dispersions of mineral particles (titanium oxide, silica, nacres, and mica).

4.6.2.4. *Examples of hair-care formulations*

There is a great variety of products in different physical forms corresponding to different objectives and needs which can vary in the case of a single use depending on the time of the year and the condition of hair.

Our purpose is not to describe all of them but to give examples of the main classes of products.

Table 4.5 gives an overview of the major technical evolutions that have taken place in the formulation of shampoos.

Shampoo 1 is representative of a "classical" shampoo. The aim is to obtain generous foam, to cleanse the hair well, without leaving it too soft while ensuring that the fibers may be untangled. It is a clear and viscous liquid.

Shampoo 2 is a "first-generation" conditioning shampoo. It is a pearlescent viscous liquid that develops abundant foam. The cationic polyelectrolyte (Polyquaternium 10) provides ease of untangling and softness, especially on wet fibers.

Shampoos 3 to 5 are representative of the latest generation of conditioning shampoos — also called "2 in 1". In example 3, the presence of a polydimethylsiloxane (Dimethicone) dispersed in the matrix of the shampoo and the presence of the cationic polyelectrolyte (polyquaternium 10) improves detangling on wet hair and softness on wet and dry hair. Example 4 is particularly well suited to damaged hair, which it protects from breakages during combing or blow-drying. In this composition, the benefits are provided by aminosiloxane dispersion, highly charged cationic polyelectrolyte (Polyquaternium 6) and the presence of ceramide (2-oleamido-1,3-octadecanediol). In example 5, the concentrations of the surfactants are adjusted to "solubilize" the cationic silicone

Table 4.5. Shampoo architectures.

INCI Name	1 (wt%)	2 (wt%)	3 (wt%)	4 (wt%)	5 (wt%)
Sodium laureth sulfate	8,5	9	14	15,5	11,2
Coco betaine	3	0	0	0	5
Disodium cocoamphodiacetate	0	3	0	0,6	0
Cocamidopropyl betaine	0	0	2,4	2,4	0
Disodium Ricinoleamido MEA-sulfosuccinate	0	0,8	0	0	0
Laureth-5 carboxylic acid	0	0	0	0	2,7
Cocamide MEA	0,5	0	0	0	0
Cocamide MIPA (and) Isopropanolamide	0	1,5	0	0	0
Cocamide MIPA (and) Methyl Cocoate (and) Sodium Cocoate	0	0	1,5	0	1,2
Polyquaternium 10	0	1	1	0	0,7
Polyquaternium 6				0,5	0
Dimethicone	0	0	2,7	0	0
Amodimethicone (and) trideceth-6 (and) cetrimonium chloride				1,6	0
Amodimethicone					1,5
2-Oleamido-1, 3-Octadecanediol	0	0	0	0,1	0
Glycol Distearate	0	1,4	0	2	0
Behenyl alcohol	0	0	1,5	0	0
Laureth 2	0	0	0,9	0	0
Distearyl ether	0	0	1,5	0	0
Carbomer	0	0,1	0,2	0,2	0
Sodium Chloride	1	2	0,6	2	0
Water, preservatives	ad. 100	ad. 100	ad. 100	ad. 100	ad. 100
pH value	6,5	7,5	7,5	5	7

(Amodimethicone). Together with the cationic polymer, this silicone provides strong conditioning properties on wet as well as on dry hair.

Examples of rinses are given in Table 4.6. Example 1 is a basic conditioning cream. Example 2 is a more sophisticated composition providing good detangling and softness properties on wet as well as on dry hair. Example 3 is a deep conditioner or masque. Its consistency allows a thick layer to be applied to the particular area where the hair is most damaged. The sorption of the components of this composition protects hair from physical damage such as erosion due to repetitive combing or blow-drying. Composition 4 is an example of a conditioning cream formulated with a more biodegradable cationic surfactant (Dipalmitoylethyl

Table 4.6.   Rinse architectures.

INCI name	1 (wt%)	2 (wt%)	3 (wt%)	4 (wt%)	5 (wt%)
Cetyl alcohol	0	3	5	0	0
Cetearyl alcohol	2, 8	0	0	6	0
Cetyl esters (and) cetyl esters	0	0, 25	1	1	0
Glyceryl stearate	0, 5	0	0	0	0
PEG-8 isostearate					4, 5
Lanolin	0, 5	0, 15	0	0	0
Mineral oil	0	0	0	2	0
Avocado oil	0	0	0	0	15
Hydroxyethylcellulose	0	0, 2	0, 2	0	0
Amodimethicone (and) trideceth-12 (and) cetrimonium chloride	0	0, 7	0	0	0
Amodimethicone (and) trideceth-6 (and) cetrimonium chloride	0	0	1, 1	1, 1	0
Amodimethicone					1, 75
Behentrimonium chloride	0	0	3		1, 5
Cetrimonium chloride	0, 55	0, 45	0	0, 8	0
Cetearyl alcohol (and) dipalmitoylethyl hydroxyethylmonium methosulfate	0	0	0	1, 35	0
Steramidopropyldimethylamine	0	0, 75	0	0	0
2-oleoamido-1,3-octadecanediol	0	0	0, 1	0	0
Lauryl PEG/PPG-18/18 methicone (and) Dodecene (and) Poloxamer 407	0	0, 25	0	0	0
PEG-180	0	2	2	0	0
Glycerin	0	0	2	0	5
Ethanol					15
Water, preservatives	ad. 100	ad. 100	ad. 100	ad. 100	ad. 100
pH value	4	5,5	3,2	3,3	5,5

hydroxyethylammonium methosulfate). Composition 5 is an example of a previously described miniemulsion.

References

1. T. Mitsui, *New Cosmetic Science Elsevier*, New York (1997).
2. F. H. C. Crick, *Acta Cryst* 6, 689 (1953).
3. R. D. B. Fraser, T. P. Macrae and E. Suzuki, *J Mol Biol* 108, 435 (1976).
4. D. A. D. Parry, *Int J Biol Macromol* l 19, 45 (1996).

5. C. Bouillon and J. Wilkinson, *The Science of Hair Care*, Taylor and Francis, New York (2005). R. Robbins, Chemical and Physical Behavior of Human Hair — 4th edition, Springer-Verlag (New York) (2002).

6. N. Yorimoto and S. Naito, *Proc Int Symp Fiber Sci Technol* Yokohama, 215 (1994).

7. P. T. Wertz and D. T. Downings, *Lipids* 23, 878 (1988).

8. P. T. Wertz and D. T. Downings, *Comp Biochim Physiol* 92B, 759 (1989).

9. D. J. Evans and M. Lanczki, *Text Res J* 67, 435 (1997).

10. C. Dauvermann-Gotsche, D. J. Evans, G. L. Corino, and A. Körner, *Proceeding of the 10th International Wool Text Res Conference*, Aachen, Germany, ST10, 1 (2000).

11. A. P. Negri, D. A. Rankin, W. G. Nelson and D. E. Rivett. *Text Res J* 66, 491 (1996).

12. A. P. Negri, D. J. Peet and D. E. Rivett, *J Text Inst* 87(1), 608 (1996).

13. S. Breakspear, J. R. Smith and G. Luengo, *J Struct Biol* 149(3), 235 (2005).

14. M. Huson, D. Evans, J. Church, S. Hutchinson, J. Maxwell and G. Corino, *J Struct Biol* 163(2), 127 (2008).

15. J. A. Swift, *J Cosmet Sci* 50, 23 (1999).

16. M. Gamez-Garcia, *J Soc Cosmet Sci* 49, 141 (1998).

17. J. A. Swift and A. C. Brown, *J Cosmet Chem* 23, 695 (1972).

18. M. L. Garcia, J. A. Epps and R. S. Yate, *J Cosmet Chem* 29, 155 (1978).

19. I. J. Kaplan, A. Schwan, and H. Zahn, *Cosmet Toil* 97, 22 (1982).

20. N. C. Kelly and V. N. E. Robinson, *J Soc Cosmet Chem* 33, 203 (1982).

21. T. Baba, N. Nagazawa, H. Ito, O. Yaida and T. Miyamoto, *Text Res J* 71, 308 (2001).

22. T. Baba, N. Nagasawa, H. Ito, O. Yaida and T. Miyamoto, *Text Res J* 71, 885 (2001).

23. V. Dupres, D. Langevin, P. Guenoun, A. Checco, G. Luengo and F. Leroy, *J Coll Interf Sci* 306(1), 34 (2006).

24. Y. K. Kamath and S. B. Ruetsch, *J Cosm Sci* 61(1), 1 (2010).

25. R. A. Lodge and B. Bhushan, *J Appl Polym Sci* 102(6), 5255 (2006).

26. R. Molina, F. Comelles, M. R. Juliá and P. Erra, *J Coll Interf Sci* 237(1), 40 (2001).

27. Y. K. Kamath, C. J. Dansizer and H. D. Weigmann, *J Appl Polym Sci* 29, 1011 (1984).

28. Y. K. Kamath, C. J. Dansizer and H. D. Weigmann, *J Appl Polym Sci* 30, 925 (1985).

29. Y. K. Kamath, C. J. Dansizer and H. D. Weigmann, *J Appl Polym Sci* 30, 937 (1985).

30. M. L. Tate, Y. K. Kamath, S. B. Ruetsch, and H. D. Weigmann, *J Soc Cosmet Chem* 44, 347 (1993).

31. D. Braida, C. Dubief, G. Lang and P. Hallegot, *Cosmet Toil* 109, 49 (1994).

32. O. Masayuki, Y. Ryoko, M. Akira, I. Shigeto, N. Shinobu, S. Satoshi, K. Emiko and S. Naoki, *J Cosmet Sci* 54, 353 (2003).

33. M. M. Breuer, G. X. Gikas and I. T. Smith, *Cosmet Toil* 94, 29 (1979).

34. F. J. Wortmann, E. Schulze zur Wiesche and A. Bierbaum, *J Cosm Sci* 54, 301 (2003).

35. N. Chamberlain and J. B. Speakman, *J Electrochem* 37, 374 (1931).

36. H. J. White and P. B. Stam, *Text Res J* 19, 136 (1949).

37. S. Carlson, *Scientific American* 6, 74 (1998).

38. M. Feughelman, *Royal Austral Chem Inst* 35(5), 2–3 (1968).

39. B. Bhushan, *Prog Mat Sci* 53(4), 585 (2008).

40. S. D. O'Connor, K. L. Komisarek, and J. D. Baldeschwieler, *J Invest Dermatol* 105, 6 (1995).

41. Y. K. Kamath and H. D. Weigmann, *J Cosm Sci* 51, 351 (2000).

42. C. R. Robbins and C. Reich, *J Soc Cosmet Chem* 37, 141 (1986).

43. C. Robbins, *J Cosm Sci* 57(3), 233 (2006).

44. H. Mizuno, M. Kjellin, N. Nordgren, T. Pettersson, V. Wallqvist, M. Fielden and M. W. Rutland, *Austral J Chem* 59(6), 390 (2006).

45. H. Mizuno, G. S. Luengo and M. Rutland *Langmuir* 26, 18909 (2010).

46. K. Ward, F. Bertails, T. Y. Kim, S. R. Marschner, M. P. Cani and M. C. Lin, *Trans Vis Comput Graph* 13, 213 (2007).

47. Y. Fukuchi and U. Tamura, *J Soc Cosmet Chem Jpn* 25, 185 (1991).

48. C. LaTorre and B. Bhushan, *Ultramicroscopy* 106(8–9), 720 (2006).

49. G. Binnig, *Phys Rev Lett* 56, 930 (1986).

50. W. Tang, H. Zhu and C.-J. Liu, *Mocaxue Xuebao/Tribology* 27(6), 588 (2007).

51. A. M. Schwartz and D. C. Knowles, *J Soc Cosmet Chem* 14, 455 (1963).

52. Y. Fukuchi, M. Okoshi and I. Murotani, *J Soc Cosmet Chem* 40, 251 (1989).

53. C. Jacques, J. Y. Kempf and A. Franbourg, *L'Oréal internal publication*, unpublished (2007).

54. H. G. Bungenberg de Jong and H. R. Kruyt, *Coll Sci Vol. 2*, Ed. Elsevier, Amsterdam (1949).

55. L. Picullel and B. Lindman, *Adv Coll Interf Sci* 41, 149 (1992).

56. K. Thalberg, B. Lindman and G. Karlsttöm, *J Phys Chem* 95, 6004 (1991).

57. Y. Wang, K. Kimura, P. L. Dubin and W. Jaeger, *Macromolecules* 33, 3324 (2000).

58. A. V. Svensson, L. Huang, E. S. Johnson, T. Nylander and L. Piculell, *Appl Mat Interf* 1(11), 2431 (2009).

59. A. V. Svensson, E. S. Johnson, T. Nylander and L. Piculell, *Appl Mat Interf* 2(1), 143 (2010).

60. J. N. Israelachvili *J Coll Interf Sci* 44, 259 (1973).

61. G. Binnig, C. F. Quate and C. H. Gerber, *Phys Rev Lett* 56, 930 (1986).

62. J. R. Smith, *J Soc Cosmet Chem* 48, 199 (1997).

63. R. L. McMullen, S. P. Kelty and J. Jachowicz, *Int J Cosmet Sci* 51, 334 (2000).

64. C. M. Mate, G. M. McCleLLand, R. Erlandsson and S. Chiang, *Phys Rev Lett* 59, 1492 (1987).

65. T. L. Phillips, T. J. Horr, M. G. Huson, P. S. Turner and R. A. Shanks, *Text Res J* 65, 445 (1995).

66. J. R. Smith and J. A. Swift, *J Microscopy* 206, 182 (2002).

67. H. You and L. Yu, *Scanning* 19, 431 (1997).
68. J. R. Smith, *J Microscopy* 191, 223 (1998).
69. R. L. McMullen and S. P. Kelty, *Scanning* 23, 337 (2001).
70. M. J. Adams, B. J. Briscoe and T. K. Wee, *J Phys D Appl Phys* 23, 406 (1990).
71. M. L. Lewis, *Amer J Orthopsych* 69(4), 504 (1999).
72. M. Sadaie, N. Nishikawa, S. Ohnishi, K. Tamada, K. Yase and M. Hara, *Colloids and Surfaces B: Biointerfaces* 51(2), 120 (2006).
73. E. D. Goddard and R. L. Schmitt, *Cosmet Toilet* 109, 55 (1994).
74. R. L. Schmitt and E. D. Goddard, *Cosmet Toilet* 109, 83 (1994).
75. V. Andre, R. Norenberg, P. Hossel and A. Pfau, *Macromolecular Symposia* 145, 169 (1999).
76. P. Hössel, R. Dieing, R. Nörenberg, A. Pfau and R. Sander, *Int J Cosm Sci* 22, 1 (2000).
77. T. Richter, J. H. Muller, U. D. Schwarz, R. Wepf and R. Wiesendanger, *Appl Phys A* 72, 125 (2001).
78. J. Blach, W. Loughlin, G. Watson and S. Myhra, *Int J Cosm Sci* 23, 165 (2001).
79. L. M. Qian, M. Charlot, E. Perez, G. Luengo, A. Potter and C. Cazeneuve, *J Phys Chem B* 108(48), 18608 (2004).
80. G. Ran, Y. Zhang, Q. Song, Y. Wang and D. Cao, *Colloids Surf B Biointerfaces* 68(1), 106 (2009).
81. S. Lee, S. Zuêrcher, A. Dorcier, G. Luengo and N. D. Spencer, *ACS Applied Materials & Interfaces* 1(9), 1938 (2009).
82. A. Charbonnelle and M. Pauly, *Drug Cosmet Ind* 9, 36 (1998).
83. J. Jachowicz, *Cosmet Toil* 113, 45 (1998).
84. H. Hoecker, *Skin Pharmacol Appl Skin Physiol* 12, 158 (1999).
85. M. V. R. Velasco, T. C. De Sá Dias, A. Z. De Freitas, N. D. V. Júnior, C. A. S. D. O. Pinto, T. M. Kaneko and A. R. Baby, *Brazilian J Pharmac Sci* 45(1), 153 (2009).
86. M. Tatsuda, M. Uemura, K. Torij and M. Matsuoka, *J Soc Chem Jpn* 21(1), 43 (1987).
87. E. Otting, M. Zimmermann and S. Hilterhauss-Bong, *J Soc Cosmet Chem* 46(2), 85 (1995).
88. T. Okumura, *Proc 4ᵗʰ Int Hair Sci Symp DWI*, Syburg (1984).
89. M. Gamez-Garcia, *J Soc Cosmet Sci* 49, 213 (1998).
90. D. Dowson, *J Engin Tribol* 223(3), 261 (2009).
91. B. Bhushan, G. Wei and P. Haddad, *Wear* 259(7–12), 1012 (2005).
92. *Standard Test Method for Abrasion Resistance of Textile Fabrics (Flexing and Abrasion Method)*, D3885-80, ASTM Book of Standards, Section 07 (1985).
93. F. Leroy, A. Franbourg, J. C. Grognet, C. Vayssie and D. Bauer, *Proc 1st Tricontinental Meet Hair Res Soc* Brussels, Belgium (1995).
94. A. Odera, *Fragrance J* 8, 82 (2002).
95. J. A. Swift, S. P. Chahal, D. L. Coulson and N. I. Challoner, *Cosmet Toilet* 116, 53 (2001).

96. J. A. Swift, D. L. Coulson and M. F. Al-Bayatti, *Proc 10th Int Wool Text Res Conf* 8, 1 (2000).
97. C. Robbins, *J Cosm Sci* 57(3), 245 (2006).
98. C. Robbins and Y. Kamath, *J Cosm Sci* 58(4), 47 (2007).
99. C. Robbins and Y. Kamath, *J Cosm Sci* 58(6), 629 (2007).
100. M. L. Garcia and J. Diaz, *J Soc Cosmet Chem* 27, 379 (1976).
101. W. Newman, G. L. Cohen and C. Hayes, *J Soc Cosmet Chem* 24, 773 (1973).
102. Y. K. Kamath and H. D. Weigmann, *J Soc Cosmet Chem* 37, 111 (1986).
103. Y. Akiyama, Y. Matsue, Y. Doi, Y. Izumi and S. Nishijima, *Nihon Reoroji Gakkaishi* 37(3), 129 (2009).
104. J. Jachowicz, G. Wis-Surel and M. L. Garcia, *J Soc Cosmet Chem* 36, 189 (1985).
105. G. Wis-Surel, J. Jachowicz and M. L. Garcia, *J Soc Cosmet Chem* 38, 341 (1987).
106. C. Dubief and V. Nardello-Rataj, *Actualité Chimique* 274, 4 (2004).
107. N. Gitis, I. Hermann, S. Kuiry and V. Khosla, *2007 Proceedings of the ASME/STLE International Joint Tribology Conference*, IJTC 2007, PART B, 939 (2008).
108. C. Jacques and A. Prieur, *L'Oréal internal publication*, unpublished (2007).
109. K. Yahagi, *J Soc Cosmet Chem* 43, 284, (1992).
110. H. M. Haake, S. Marten, W. Seipel and W. Eisfeld, *J Cosmet Sci* 60(2), 143 (2009).
111. G. M. Eccleston and A. T. Florence, *Int J Cosmet Sci* 7, 195 (1985).
112. G. M. Eccleston, *J Coll Interf Sci* 57, 66 (1976).
113. B. W. Barry and G. M. Saunders, *J Coll Interf Sci* 41, 331 (1972).
114. B. W. Barry and G. M. Saunders, *J Coll Intef Sci* 34, 300 (1970).
115. B. W. Barry and G. M. Saunders, *J Coll Inerf Sci* 36, 130 (1971).
116. EP227994 July 8, 1987 (assigned to the Kao Co).
117. EP495624 July 22, 1992 (assigned to the Unilever Co).
118. EP500437 August 26, 1992 (assigned to l'Oréal).
119. B. A. Bernard, A. Franbourg, A. M. François, B. Gautier and P. Hallegot, *Int J Cosmet Sci* 24, 1 (2002).
120. M. D. Berthiaume and J. Jachowicz, *J Coll Interf Sci* 141, 299 (1991).
121. J. Jachowicz and M. D. Berthiaume, *J Coll Interf Sci* 133, 118 (1989).
122. S. Watanabe and K. Yahagi, *Nihon Keshôhin Gijutsushakai-Shi* 29(1), 64 (1995).
123. T. V. Drovetskaya, E. F. Diantonio and R. L. Kreeger, *J Cosm Sci* 58, 421 (2007).
124. S. Rogasik, N. Martin, J. M. Ricca, W. Wielinga and O. Anthony, *SOFW-J* 125, 32 (1999).
125. M. Gamez-Garcia, *IFSCC Magazine* 4, 99 (2001).
126. EP181773 May 21, 1986 (assigned to the Procter & Gamble Co).
127. EP 190010, August 6, 1986 (assigned to the Procter & Gamble Co).
128. M. Gamez-Garcia, J. V. Gruber, F. Winnik and B. R. Lamouroureux, *XXIst IFSCC Int. Congress Proceedings* p. 176, Berlin (2000).

Chapter 5

Hydrogel Friction and Lubrication

Jian Liu

Graduate School of Science Hokkaido University
Kita-10-Nishi-8, Kita-Ku Sapporo, 060-0810, Japan

Jian Ping Gong

Faculty of Advanced Life Science Hokkaido University
Kita-10-Nishi-8, Kita-Ku Sapporo, 060-0810, Japan
gong@mail.sci.hokudai.ac.jp

5.1. Introduction

Sliding friction is one of the oldest problems in physics and certainly one of the most important subjects from a practical point of view. Many different terms and approaches have been used to describe friction.[1-3] For friction in the presence of lubricating oil, the behavior is usually divided into three regimes. In the boundary lubrication (BL) regime, which occurs at low sliding velocity when there is negligible fluid entrainment into the contact zone, the load is carried by the contacting asperities and friction is dependent on the surface and interfacial film properties at the molecular scale. In the hydrodynamic lubrication regime, a film of lubricant, whose thickness depends on the viscosity and entrainment velocity, is entrained to fully separate the solid surfaces. The friction now depends on the rheological properties of the lubricant film in the contact zone, under the high-shear-rate condition that prevails there. Sophisticated solutions of the Reynolds equation have evolved to explain friction on the basis of the continuum flow of liquids. The fundamental view here is that liquids resist deformation against force that increases with velocity.[3] The dependence of force on velocity is linear for the simplest case (said to be "Newtonian"). Even modern treatments consider that when velocity is increased, the force required to accomplish deformation increases with a positive slope. The mixed regime is termed as a friction transition regime here, which lies between boundary lubrication and hydrodynamic lubrication. In this regime both the boundary film and bulk lubricant play crucial roles in determining friction.

Some biological surfaces display fascinating low-friction properties.[4–13] For example, the cartilage in animal joints displays a friction coefficient in the range 0.001–0.03 — remarkably low even for hydrodynamically lubricated journal bearings.[4,5] It is not well understood why the cartilage friction of the joints is so low even under conditions in which the pressure between the bone surfaces reaches values as high as 3–18 MPa and the sliding velocity is never greater than a few centimeters per second.[4] Under such conditions, it is considered that the lubricating liquid layer cannot be sustained between two solid surfaces and hydrodynamic lubrication, in the usual sense, does not work. The authors consider that these fascinating tribological properties of biological systems originate from the soft and wet nature of tissues and organs. Cartilage cells synthesize a complex extracellular matrix (ECM); the weight-bearing and lubrication properties of cartilage are associated primarily with this matrix and its high water content (ca. 75–80%). The proteoglycan, aggrecan, and the cross-linked network of collagen fibrils, the main macromolecular constituents of ECM solvated by water,[9,10] exist as a gel state, which is critically important for the specific frictional properties of the biological systems.

As soft and wet materials, hydrogels look like a solid material but are capable of undergoing large deformations. A hydrogel consists of cross-linked hydrophilic polymer networks solvated with water. Hydrogels display properties characteristic of both solids and liquids. Like solids, they deform with stress and recover their initial shape after removal of the stress. Like liquids, they can support fluid convection and diffusion of solutes that are smaller than the mesh size of the network. The highly deformable double network (DN) gel has 90% water content and is one of the most novel hydrogels (Fig. 5.1).[14] Studying the friction behavior of hydrogels is helpful for understanding the low-friction mechanism observed in biological systems, and may be useful in finding novel approaches in the design of low-friction artificial organs.

In order to elucidate the general tribological features of a solvated polymer matrix, friction of various kinds of hydrogels under different conditions has been investigated over the last several years, and very rich and complex frictional behavior has been observed.[15–34] To describe the frictional behavior of a gel sliding against a solid countersurface, we have proposed a thermodynamic model (Repulsion and Adsorption Model) from the viewpoint of

Fig. 5.1. Photos of double network (DN) hydrogel sustaining high compression. (Reproduced with permission from Ref. 67).

polymer-solid interfacial interaction.[17] Based on the repulsion and adsorption model, we can also understand the frictional properties of gel sliding against gel.[18,56]

This chapter consists of five sections that summarize the experimental and theoretical results on the surface friction and lubrication of hydrogels. In Sec. 5.3, the repulsion and adsorption model of gel friction will be discussed, to facilitate understanding the friction of hydrogels. In Sec. 5.4, the frictional properties of neutral hydrogels are described. In Sec. 5.5, the frictional properties of polyelectrolyte hydrogels will be discussed in comparison with the previous section. In Sec. 5.6, we will deal with the low friction of surface-modified hydrogels. In Sec. 5.7, attempts to apply the robust, low-friction double-network hydrogels as substitutes for biological tissues such as artificial cartilage are discussed.

5.2. Experimental Details

5.2.1. *Sample preparation*

Physically cross-linked poly (vinyl alcohol) PVA gel was prepared by a freezing ($-40°C$) and thawing ($25°C$) method from an appropriate PVA aqueous ($DMSO:H_2O = 67.5 : 22.5$ w/w) solution ($\sim$10 wt %).[70] Solutions were prepared by heating PVA in the aqueous medium for 1 h at $\sim$90°C. After the heating step, the PVA solutions were placed in a dryer connected with a vacuum pump, to facilitate the release of air bubbles. When all air bubbles were removed, the solutions were cast between glass plates and quenched for 24 h at $-40°C$. Following the quenching period, hydrogel sheets ($\sim$3 mm thick) were allowed to warm up to room temperature and then submerged in copious distilled water for at least one week to extract DMSO.

The polyelectrolyte gel, poly(2-acrylamido-2-methyl-1-propanesulfonic acid) (PAMPS) gel was synthesized by radical polymerization. An aqueous solution of 1.0 M 2-acrylamido-2-methyl-1-propanesulfonic acid (AMPS), 5.0 mol% cross-linking agent, N, N'-methylenebisacrylamide (MBAA), and 0.1 mol% initiator, 2-oxoglutaric acid in the reaction cell was purged with nitrogen gas for 30–45 min and irradiated with UV light for 7 h at 20°C. The reaction was carried out between a pair of glass substrates, separated with a 2-mm-thick spacer. After polymerization, the gel was immersed in a large amount of water for 1 week to equilibrate and to wash away the residual chemicals. The surface of gel prepared on the glass was mirror-like and was used for measurements. Other samples were prepared in the same manner and details have been described in Refs. 18–24.

5.2.2. *Measurements*

5.2.2.1. *Friction measurement using rheometer*

A commercial rheometer (3-ARES-17A, Rheometric Scientific Inc. USA), as shown in Fig. 5.2, was used for measuring the friction of a gel against a solid surface or a gel in water or in aqueous NaCl solutions at prescribed temperatures. Samples were glued to the upper surface of a coaxial disk-shaped platen. As the opposing countersurface, a gel or glass plate a little larger than the upper gel was fixed on the lower platen. The interface was immersed in water or in aqueous salt solutions.

5.2.2.2. *Friction measurement using tribometer*

The friction of gels against a piece of solid countersurface such as a piece of glass was also measured in air or in water using a commercial tribometer (Heidon 14S/14DR, SHINTO Scientific Co Ltd, Japan), as shown in Fig. 5.3. A gel sample with a prescribed size was embedded in a square frame of adjustable size attached to the upper plate and pressed against a piece of glass, which was fixed on the lower plate and was driven to move horizontally and repeatedly at a prescribed velocity over a distance of 90 mm.

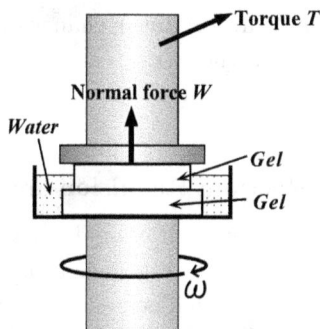

Fig. 5.2.   Schematic illustration of the rheometer used to measure the gel friction. Reproduced with permission from Ref. 18. Copyright © 1999, Am Chem Soc.

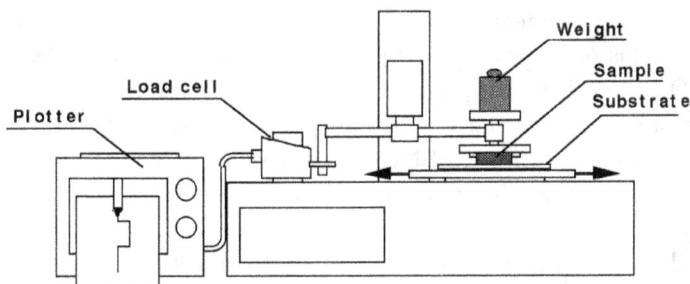

Fig. 5.3.   Illustration of the tribometer used to measure the gel friction. Reproduced with permission from Ref. 19. Copyright © 1999, *Am Chem Soc.*

The normal load (or sliding velocity) dependence was studied by using a sample for which the load (or velocity) was changed in a stepwise manner from lower to higher values without separating the two surfaces during the interval of measurement. A detailed description of the measurements has been given in Ref. 19.

The difference in the measurements between using a tribometer and a rheometer is that with the former, the friction tests are performed under a constant normal compressive stress, but the latter runs in a constant compressive strain mode. In both cases, prior to the measurement, the gel sample was loaded with a normal load (or normal stress) and reached an equilibrium state. After achieving relaxation equilibrium, a displacement with a velocity, $v(\omega)$, was applied to the lower platen of the tribometer (rheometer) to generate the frictional force (torque). By using a rheometer, the total frictional force, F can be related to the torque by assuring that the unknown frictional shear at a radius r changes with the sliding velocity in a power law as $F \propto (\omega r)^{\alpha}$, where α is a constant. A detailed description of measurement has been given in Ref. 18.

5.3. A Model of Gel Friction: Repulsion and Adsorption

As polymer gels have very complex frictional behaviors, which will be presented later, we would first like to introduce a primary concept about gel friction. This may make the frictional behaviors of gels easier to understand. Based on the studies of gel friction, Gong and co-workers have proposed a repulsion-adsorption model from the friction of polymer-solid interfacial interactions to describe the friction behavior of gels against a smooth countersurface, as shown in Fig. 5.4.[17]

A gel has many features in common with other soft cross-linked polymers in terms of its cross-linked polymer structure and viscoelastic properties. On the other hand, in contrast to a rubber, a gel contains a large amount of a low molecular weight component — water. The water content in biological gels, such as cartilage and other soft tissues, is *ca.* 70–80 wt% and in synthetic gels, the water content can be as high as 99.9 wt%. Water in gels is strongly solvated to the polymer network and cannot be squeezed out easily like a sponge. Due to the presence of a large amount of water,

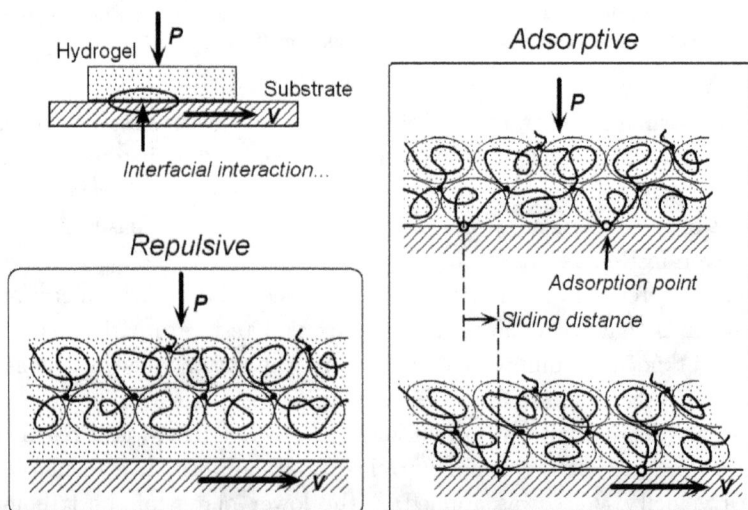

Fig. 5.4. Schematic illustration of the repulsion-adsorption model for gel friction against a solid countersurface. Reproduced with permission from Ref. 15.

the internal friction of a gel is much less than that of a cross-linked polymer melt or rubber. For a typical gel, $\tan \delta$, a characteristic parameter of the viscoelastic properties of soft materials, is in the order of 0.01, while that of a rubber is never less than 0.1.

The most simple molecular picture of a polymer gel is given by a C* gel in scaling theory.[35] Scaling theory describes a gel as a collection of adjacent blobs with radius R_F and having a characteristic relaxation time $\tau_f \approx \eta R_F^3 / T$, where T is the absolute temperature in energy units, and η is the viscosity of the solvent. Each blob is associated with one partial polymer chain (the polymer chain between two next neighboring cross-linking points). The scaling theory relates this molecular structure of the gel with its elastic modulus E by $E \cong T / R_F^3$.[35]

The main argument of the adsorption-repulsion model is as follows: Considering a C* gel in contact with a solid wall in water, in analogy to a polymer solution, the polymer network on the surface of the gel will be repelled from the solid surface if the interface interaction is repulsive, and will be adsorbed to the solid surface if it is attractive.

In the repulsive case, the friction is due to the lubrication of the hydrated water layer of the polymer network at the interface, which predicts that the frictional stress σ should be proportional to the sliding velocity.

In the attractive case, the friction of a gel is from two contributions: (1) elastic deformation of adsorbing polymer chain, (2) lubrication of the hydrated layer of the polymer network, where the first contribution is the same as the adhesive friction as proposed by Schallamach.[36–38]

According to Schallamach's model for rubber,[36] the frictional stress from the first contribution, σ_{el}, arising from the elastic stretching of the polymer chains adsorbed on the countersurface, is expressed as the product of the adsorbing site number per unit area, m, the stretching velocity, ν, and the life-time of adsorption, τ_b, that is, $\sigma_{el} \propto m\nu\tau_b$. It takes a time τ_f for a desorbed polymer to be re-adsorbed on the surface. Taking into account the thermal agitation, $\tau_b^{-1} = \tau_f^{-1} \exp[-(F_{ads} - F_{el})/T]$. Here, F_{ads} is the adsorption energy per polymer chain, and F_{el} is the elastic energy due to stretching of the adsorbing chain. The elastic energy accumulated by stretching favors the adsorption. Therefore, τ_b decreases with an increase in chain stretching.

By applying the scaling relation to Schallamach's model, one obtains a characteristic velocity[17]:

$$\nu_f \approx R_F / \tau_f = T / \eta R_F^2 \tag{1}$$

and it can be expressed in terms of elastic modulus E of the gel,

$$\nu_f \approx T^{1/3} E^{2/3} / \eta \tag{2}$$

σ_{el} has a maximum at $\nu \tau_f / R_F \approx 1$. When the sliding velocity is slow enough so that $\nu \tau_f / R_F \ll 1$, the friction force is due to the elastic force of stretched polymer chain, and it increases with the sliding velocity. When the sliding velocity is high so that $\nu \tau_f / R_F \gg 1$, the polymer has insufficient time to form an adsorbing site, and the friction decreases with the increase of the velocity in this velocity region. This characteristic peak of elastic friction is related to the viscoelastic G"-peak of the gel since the latter is also determined by the characteristic relaxation time of the network τ_f and mesh size R_F.

The friction from the second contribution, that is, σ_{vis} the viscous friction, increases with the velocity monotonically. Therefore, the first contribution is dominant when $\nu \tau_f / R_F \ll 1$, and the second contribution is dominant when $\nu \tau_f / R_F \gg 1$. Around $\nu \tau_f / R_F \approx 1$, a transition from elastic friction to hydrated layer lubrication occurs, as shown schematically in Fig. 5.5.

The repulsion–adsorption model also predicts that: (1) when the countersurface is repulsive, the frictional force is lower than in the attractive case, and it linearly increases with the normal pressure and velocity when the compressive strain (normal load) is not very high, (2) when the countersurface is attractive, the frictional force increases with the attraction

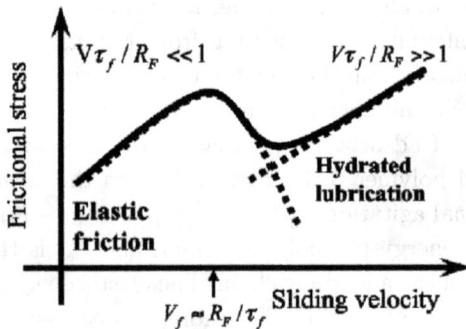

Fig. 5.5.   Schematic curve for the friction of a gel that is adhesive to the countersurface in liquid. The friction is the sum of elastic force due to polymer adsorption and viscous force due to hydration of polymer. At $v \ll v_f$, the first component is dominant. At $v \gg v_f$, the second component is dominant. Transition from elastic friction to lubrication occurs at the sliding velocity characterized by the polymer chain dynamics $v_f \approx R_F / \tau_f = T / \eta R_F^2$. Reproduced with permission from Ref. 30.

strength. For weak attraction, the pressure dependence of the frictional force is much weaker than in the repulsive case. It becomes stronger when the attraction strength increases.[17]

In the following section, we will present some experimental results of gel friction, in most studies against glass countersurfaces, using a poly (vinyl alcohol) (PVA) hydrogel as the attractive case and poly (2-acrylamido-2-methyl-1-propanesulfonic acid) (PAMPS) hydrogel as the repulsive case. We should note that beside their different interactions with the countersurfaces, another important difference between the neutral PVA gel and the charged PAMPS gel is their different elastic modulus. Even for the same amount of water content, the neutral PVA gel is softer than the negatively charged polyelectrolyte gel PAMPS, which will also influence the frictional behavior.

5.4. Frictional Properties of a Neutral Hydrogel: PVA Gel

5.4.1. *PVA gel on smooth glass: effect of temperature*

The repulsion-adsorption model implies that, based on Eqns. (1) and (2), the adhesive friction is determined by the scaled $\nu\tau_f/R_F$, which depends on (1) elastic modulus E or R_F, the latter being related to the partial chain length between the neighboring cross-linking points as well as on the solvent quality; (2) viscosity of solvent; (3) temperature.

PVA gel immersed in water has a weak attractive interaction with a glass surface, as confirmed by AFM.[19] Figure 5.6 shows the frictional forces when a physically cross-linked PVA gel is slid against a glass surface in water at room temperature.[22]

The friction increases with the sliding velocity and reaches a maximum at an average sliding velocity of 3.8×10^{-2} ms^{-1}, which agrees with theoretical predictions.[17] From the velocity at which the frictional force attains a maximum, the radius of the partial polymer chain, R_F, was estimated as 16 nm by using Eq. (1). On the other hand, the Flory radius of the polymer chain R_F was about 3 nm, estimated from the swelling degree $q = 11$ using the $C*$ gel theory. Considering that these order estimations are made using scaling relations, in which all the numerical factors are neglected, these two results are in good agreement.

The temperature effect on the gel friction is shown in Fig. 5.7.[22] An increase in temperature from 5°C to 45°C leads to both a decrease in the frictional force and an increase in the velocity where friction force shows the maximum. This temperature dependence agrees well with the

J. Liu and J. P. Gong

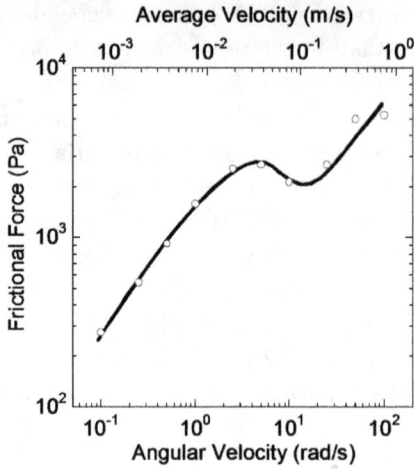

Fig. 5.6. Velocity dependence of the frictional force for ring-shaped PVA gel rotated on a glass surface in pure water at 25°C measured with a rheometer. Sample size; inner radius: 5 mm, outer radius: 10 mm, thickness: 3 mm; degree of swelling q(g/g): 11; normal strain: 30%; normal stress: 8.5 kPa. Reproduced with permission from Ref. 22.

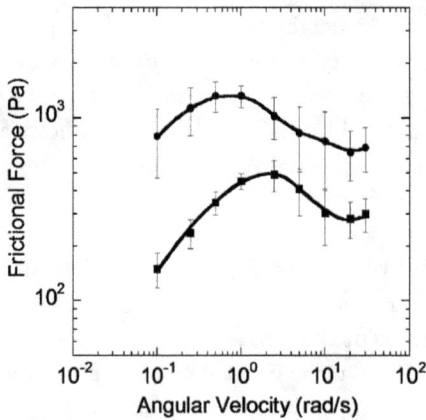

Fig. 5.7. Frictional forces as a function of the angular velocity for ring-shaped PVA gel rotated against a glass surface in pure water at 5°C (●) and 45°C (■) measured by a rheometer. Sample size; inner radius: 5 mm, outer radius: 10 mm, thickness: 3 mm; degree of swelling: 11. The error ranges in the figure are standard deviations of the mean values over 4 samples. Reproduced with permission from Ref. 22.

surface-adhesion mechanism: An increase in temperature should result in a decreased friction force, due to an increase in the thermal agitation that favors desorption. At the same time, ν_f of the polymer chain increases with an increase in temperature, originating partly from the increased thermal energy and partly from the decreased viscosity of the solvent, as shown by Eq. (1). As shown in Fig. 5.7, when the temperature is raised from 5°C to 45°C, ν_f increases 5 times. The theory (Eq. (1)) predicts about 3 times increase in ν_f, which roughly agrees with the experimental results.

5.4.2. *Countersurface effects*

The frictional behavior of gels should also depend against the opposite countersurface as we discussed in the section of the repulsion-adsorption model. The frictional behaviors of PVA gel against smooth countersurfaces with various levels of hydrophobicity and against rough countersurfaces with different roughness have been studied.[33,34]

5.4.2.1. *Countersurface adhesion and hydrophobicity*

Figure 5.8 shows the velocity dependence of the PVA gel friction against countersurfaces with different level of hydrophobicity.[33]

As shown in Fig. 5.8, the friction behavior of hydrogels sliding against smooth countersurfaces strongly depends on the adhesion strength and hydrophobicity of the countersurfaces. On a hydrophilic countersurface that

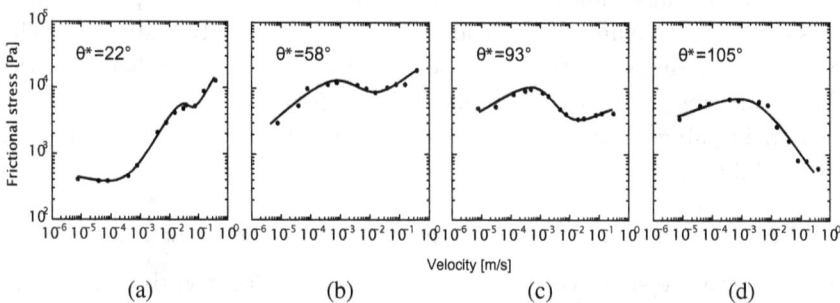

Fig. 5.8. Velocity dependence of dynamic frictional stress of PVA gel against solid countersurfaces of various hydrophobicities in water. (a) G2; (b) SW; (c) OTS-glass; and (d) F-glass. Hydrophobicity of the countersurfaces is in the order G2 < SW < OTS − glass < F − glass. Normal strain: 26%; normal stress: $14\,kPa \cdot \theta^*$ is the contact angle. Reproduced with permission from Ref. 33.

is weakly adhesive (contact angle $\theta^* = 22°$) the friction-velocity relation is divided into three regions: a creep region at low velocity, an elastic region at intermediate velocities, and a lubricating region at high velocities. The friction behavior is satisfactorily described by the adsorption model described in the previous section except the creep region.[33] There is no distinct water invasion by the so-called hydrodynamic effect, and the lubricating layer is formed by the polymer hydration with a thickness of the order of ξ, which is velocity-independent.

Against strongly adhesive countersurfaces ($\theta^* = 58°$, 93°), friction increases with the countersurface hydrophobicity in the low-velocity region, showing a weak velocity-strengthening, and a dramatic friction transition occurs at around $\nu = 10^{-3}$ m/s; this value is one order lower than the characteristic velocity of the polymer chain $\nu_f = \xi/\tau_f$. The friction behavior is satisfactorily described by the adsorption model below the transition region, while this friction transition is explained in terms of the elastic dewetting-wetting transition that is caused by the invasion of trapped water formed due to the heterogeneous dewetting.[33]

Furthermore, when the countersurface has a contact angle higher than 105°, the friction exhibits a bell-shape, which monotonously drops at $\nu > 10^{-3}$ m/s. On such a strongly hydrophobic surface, the apparent lubrication layer thickness is estimated to be on the order of several hundred nanometers. Formation of such a thick lubricating layer cannot be satisfactorily explained in terms of the dewetting-wetting transition model, and nano-bubble layer formation on the hydrophobic countersurface might account for this interesting phenomenon.[44–52]

Figure 5.8 also indicates that even against the most hydrophobic countersurface, the maximum frictional stress is approximately $1/10 \sim 1/5$ of the gel elastic modulus under a large normal strain of 26%. Furthermore, the frictional stress against a hydrophobic countersurface in the high-velocity region, i.e. above the transition velocity, is much lower than that on a hydrophilic surface.

5.4.2.2. *Friction of a soft hydrogel against rough solid countersurfaces*

The velocity dependence of friction on rough surfaces is determined by surface contact dynamics and characterized by two velocities, i.e. ν_f and $\nu_{drainage}$. The former is determined by the cooperative diffusion constant of the gels, and the latter by the surface roughness of the countersurface and the normal pressure applied on the gel.[34]

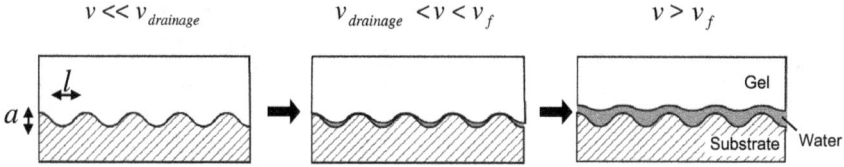

Fig. 5.9. Schematic illustration of the contact of a soft gel sliding against a rough solid countersurface. The countersurface roughness l is assumed to be much larger than the mesh size ξ of the gel. As the sliding velocity increases, the effective contact area decreases. Reproduced with permission from Ref. 34.

Against a rough countersurface of characteristic asperity wavelength l and height a, the effect of sliding velocity on the contact area between the gel and countersurface should be considered. Figure 5.9 shows a schematic drawing of the dependence of the sliding velocity on the contact in the case of a soft gel against a rough solid countersurface in water.

The effective contact area is influenced by the drainage time of water entrapped between the gel and countersurface. Therefore, when $v \ll v_{drainage}$, σ_{el} (friction from stretching the adsorbed polymer chains) increases with an increase in roughness due to enhancement of the real surface contact area. When $v_{drainage} < v < v_f$, the gel does not have sufficient time to form contacts on asperities, and the effective contact area dramatically decreases. Thus, σ_{el} decreases as the velocity increases. At $v > v_f$, the polymer is incapable of being adsorbed on the countersurface, and lubrication becomes dominant. Consequently, σ_{el} becomes less important than σ_{vis} (friction from hydrodynamic lubrication). Since the viscous term σ_{vis} increases with the velocity, we again expect an increase in friction with velocity in this regime.

We should note that the energy dissipation of the bulk gel near the countersurface asperities due to mechanical vibration caused by sliding might not be significant.[34] This is because a hydrogel has a very low internal friction between polymer chains due to the presence of a large amount of water. This is different from the case of a rubber, in which the bulk internal friction should also dissipate a significant amount of energy when being slid on a rough surface.[12,53]

Figure 5.10 shows the velocity dependence of the PVA gel on soft glass and BK7 glass of different roughness. When the countersurface roughness increases, the friction in the low-velocity region slightly increases, while it does not change remarkably at high velocity when the countersurface roughness R_a is below 100 nm. For a rougher surface with $R_a = 1000$ nm,

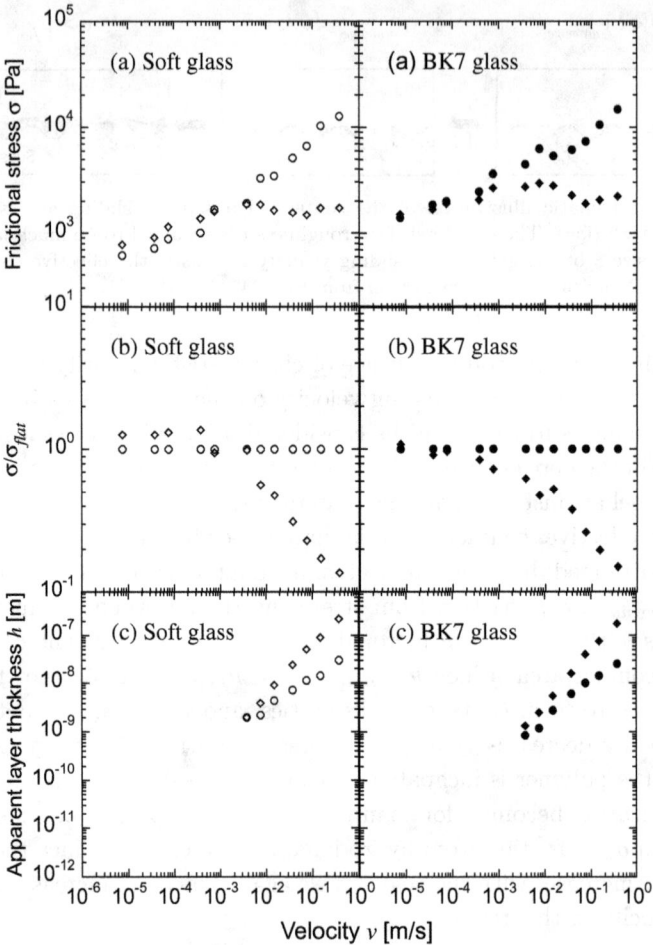

Fig. 5.10.   Velocity dependence of dynamic frictional stress (a), dynamic frictional stress normalized by the value on a flat surface (G20 or BK7 1/10 λ) (b), the apparent water layer thickness, estimated by $h = \eta v / \sigma$ (c), of the PVA2000 gel sliding against soft glass (left column, open symbols) and BK7 glass (right column, closed symbols) of different roughness. The measurements were performed in water at 20°C under a normal strain of 26%. (○): G20, $R_a = 20$ nm; (◇): G1000, $R_a = 1000$ nm. (●): BK7 1/10 λ, $R_a = 20$ nm; (◆): BK7 frost, $R_a = 1000$ nm. Reproduced with permission from Ref. 34.

the friction begins to decrease slightly at a sliding velocity of $v = 10^{-4} - 10^{-3}$ m/s and abruptly at $v \approx 10^{-2}$ m/s. When $v > 10^{-1}$ m/s, the friction shows a tendency to increase again with an increase in velocity. This behavior can be explained in terms of contact dynamics, which features two characteristic velocities, that is, ν_f and $\nu_{drainage}$. When the characteristic length of asperities is much larger than the mesh size of the gel ($l \geq a \gg \xi$), we have $\nu_{drainage} \ll \nu_f$. At $\nu < \nu_{drainage}$, the friction increases with the roughness due to the enhanced contact area between the gel and the countersurface. At $\nu_{drainage} < \nu < \nu_f$, the gel does not have sufficient time to form complete contacts with the surface asperities and the contact area decreases, which leads to a decrease in friction with the velocity. At $\nu \gg \nu_f$, there is no contact between the gel and countersurface, and a continuous lubricating layer forms at the interface. This again leads to an increase in the frictional stress with velocity.[34]

5.4.3. *Polymer solution as lubricant*

The friction events in biological systems mostly occur between soft and wet tissues mediated by viscoelastic polymer fluids, such as synovial fluid or mucus.[4,5] For example, mucus adheres to many epithelial surfaces, where it serves as a diffusion barrier against contact with noxious substances (e.g. gastric acid, smoke) and as a lubricant to minimize shear stresses; such mucus coatings are particularly prominent on the epithelia of the respiratory, gastrointestinal and genital tracts. Mucus is also an abundant and important component of saliva, giving it virtually unparalleled lubricating properties.[4,5]

The effect of polyethylene oxide (PEO) solution on the friction of a PVA gel against glass has been studied.[54] The previous section showed that the surface friction of PVA gels against smooth, adhesive glass countersurface derives from two contributions: namely surface adhesion and hydrated lubrication. The former is dominant at low sliding velocities and the latter at high velocities. When a water-swollen PVA gel is pressed onto a glass surface in dilute ($c/c^* \leq 1$) PEO aqueous solution, a layer of PEO coils is confined in the space between PVA gel and glass surface, in which the PEO concentration is much higher than that of bulk solution for very long PEO (PEO 4E6). Figure 5.11 gives the friction behavior of PVA gel against a glass surface in dilute PEO solutions with molecular weight of 2×10^4 (PEO 2E4).[54]

When the actual PEO concentration in the confined space is lower than the overlap concentration, PEO blobs screen the adsorption of PVA blobs to the glass countersurface, and this leads to a lower frictional stress than that in pure water in the slow, sliding region (PEO 2E4 $c/c^* \leq 1$). This screening effect is found to be insensitive to normal pressure.[54]

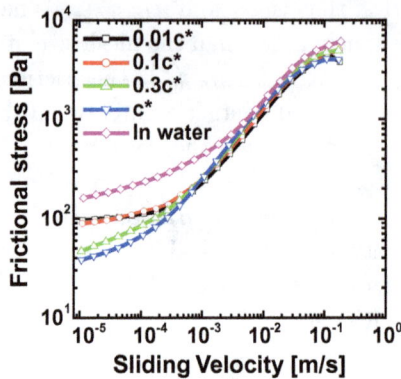

Fig. 5.11.   Sliding-velocity dependence of the frictional stress for PVA gel sliding against a glass surface in dilute PEO 2E4 solution with different concentrations (relative to the polymer overlap concentration, c*) at 20°C. Sample thickness: ∼3 mm. Normal pressure: 14 kPa. Reproduced with permission from Ref. 54.

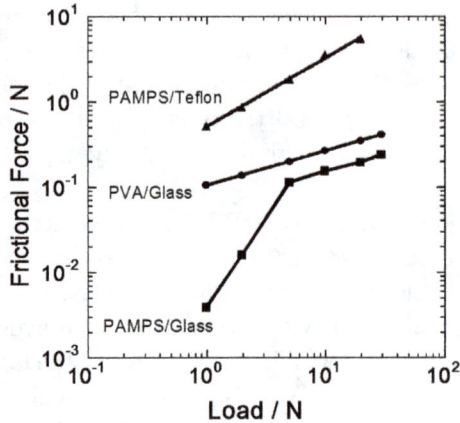

Fig. 5.12.   Load dependence of gel friction in water measured by a tribometer. Reproduced with permission from Ref. 55.

5.5. Frictional Properties of Polyelectrolyte Hydrogels

5.5.1. *Friction of gels against countersurfaces: effect of interfacial interaction*

The frictional behavior of gels depends against the countersurfaces. When the non-ionic PVA gel is allowed to slide on a polytetrafluoroethylene (Teflon) plate, for example, the behavior is the same as that on a glass

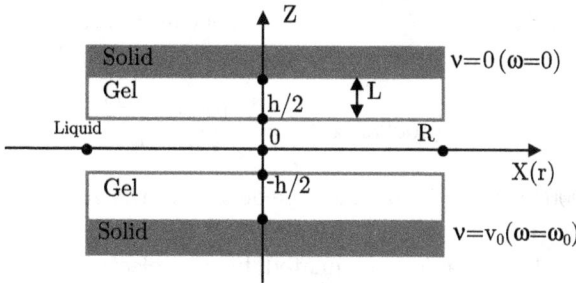

Fig. 5.13. The geometry of two like-charge gels approaching each other in water. Due to the osmotic repulsion exerted by the dissociated counter-ions, a water layer h in thickness is formed under a normal pressure P. One of the gels is undergoing translational (or rotational) motion at a velocity of ν_0 (or ω). Reproduced with permission from Ref. 18.

surface. However, the behavior of strongly anionic PAMPS gel on Teflon greatly changes and becomes similar to that of PVA on glass, as shown in Fig. 5.12. When a pair of polyelectrolyte gels carrying the same charges, for example PNaAMPS gel with PNaAMPS gel, are slid against each other, very low frictional force is observed.[19] On the other hand, when two polyelectrolyte gels carrying opposite charges are slid against each other, the adhesion between the two gels is so high that the gels are broken during the measurement.[18] The phenomenon indicates that the interfacial interaction between the gel surface and the countersurface is crucial in gel friction.

AFM results have shown that PAMPS is repulsive to a glass surface and adhesive to a Teflon surface, while PVA is attractive to glass. These results also confirm that the sliding friction of gels is closely related to polymer chain-countersurface interactions and supports the theoretical prediction that the friction of a gel is dominated by the hydrodynamic lubrication mechanism when the gel-countersurface interaction is repulsive and by the adhesion mechanism when the interaction is attractive.[55]

5.5.2. *Friction between two like-charged gels*

5.5.2.1. *Effect of charge density and electrical double layer*

According to the repulsion-adsorption model, the low friction between two polyelectrolyte gels carrying the same charge involves a hydrodynamic mechanism, in which the strong osmotic repulsion between the dissociated counter-ions of the two surfaces sustains a solvent layer between two gel surfaces against a normal pressure, as shown in Fig. 5.13.

Supposing that the polyions are homogeneously distributed in the bulk gel as well as on the gel surface, then the average surface charge density is[18]

$$\sigma = (1000cN_A)^{2/3} = \left(\frac{10^6 N_A}{qM_W}\right)^{2/3} \tag{3}$$

Here, $c = 1000/qM_w$ is the bulk charge density of the gel or the monomeric concentration (in units of mol/L) and N_A is the Avogadro's number. The solvent layer thickness h is estimated by the electrostatic double layer theory.[18]

$$h = 2\sqrt{2kT/\mathrm{P}r_0}\arctan(\sigma\sqrt{kTr_0/2P}) \tag{4}$$

Here, k is the Boltzmann constant, T is the absolute temperature, $r_0 = e^2/\varepsilon kT$ is a constant with a dimension of length, and P is the applied pressure.

As in the repulsive case, the friction is determined by the shear stress at the gel surface[18]:

$$\sigma_{vis} = \frac{\eta\nu_0}{2(h/2 + \sqrt{K_{gel}})} \tag{5}$$

Here $\sqrt{K_{gel}} \cong \xi$ decreases with the increase in the polymer network density. So the friction coefficient for disc-shaped gels of radius R is[18]:

$$\mu = \frac{F}{\pi R^2 P} = \int_0^R 2\pi r \sigma_{vis} dr = \frac{\eta\omega_0 R}{3P(h/2 + \xi)} \tag{6}$$

Figure 5.14 shows the network charge density dependence of the friction for two pieces of PAMPS gels undergoing relative rotation. The modulation of charge density of the gel has been achieved by varying the amount of cross-linking agent during the process of gel synthesis, which gives rise to gels of different swelling ability. Upon increasing the charge density, the repulsion between two gels will be increased, which favors the formation of a thicker solvent layer and decrease the viscous friction. At the same time, upon increasing the network density the mesh size of the gels will decrease, which increases the friction. As shown in Fig. 5.14, the friction increases modestly with the increase in charge density, due to these two opposite effects.

The effect of charge density was also studied by sliding copolymer gels containing different numbers of ionic monomers on charged homopolymer gels. Figure 5.15 shows the friction force when poly(AAm-co-NaAMPS) gels with various composition were allowed to slide on a PNaAMPS gel.

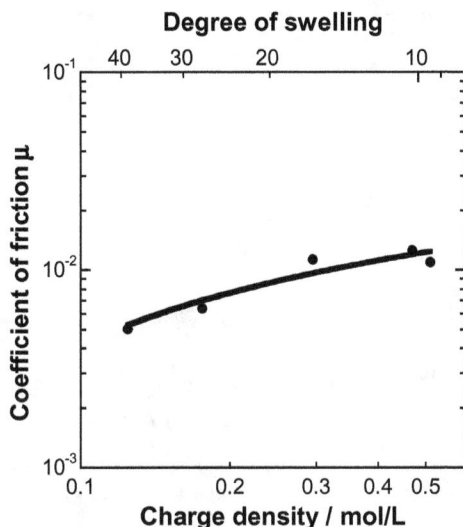

Fig. 5.14. Charge density and degree of swelling dependencies of friction coefficient between two PAMPS gels in water. Sample radius R = 7.5 mm; angular velocity ω = 0.05 rad/s. Initial normal force W = 3 N. Reproduced with permission from Ref. 18.

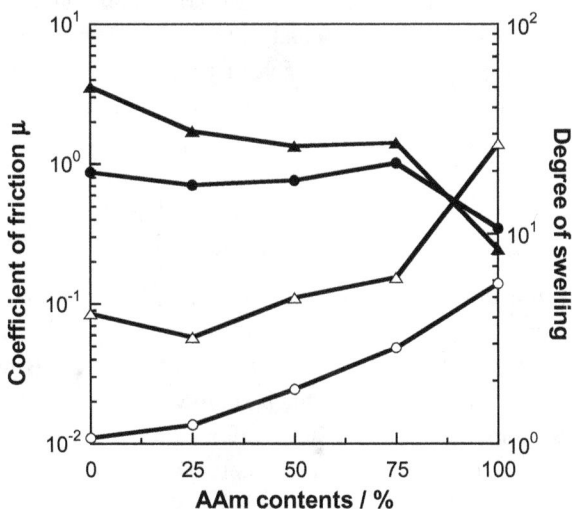

Fig. 5.15. Dependencies of the friction coefficient (open symbols) and the degree of swelling (solid symbols) on the copolymer composition. The friction values are between poly(AAm-co-NaAMPS) gel and PNaAMPS gel (○, ●) or poly(AAm-co-DMAPAA-Q) gel and PDMAPAA-Q gel (△, ▲) in water. Sample radius R = 7.5 mm; angular velocity ω = 0.01 rad/s. Initial normal force W = 3 N. Reproduced with permission from Ref. 18.

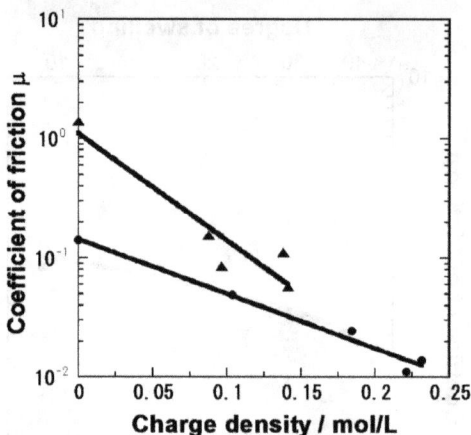

Fig. 5.16. Relations between the friction coefficient and the gel charge concentration calculated from Fig. 5.7. (●) poly(AAm-co-NaAMPS) gels; (▲) poly(AAm-co-DMAPAA-Q) gels. Reproduced with permission from Ref. 18.

According to the theory above, with increased charge density and without changing the swelling degree, Figure 5.16 clearly shows that the coefficient of friction decreases with the increase in the charge density.

From the above results, we could demonstrate that the friction of gels is largely dependent on the charge density of the gels. By varying the charge density of the gel or the ionic strength of the solvent, the friction coefficient of the gel can be varied by about two orders of the magnitude in the examined experimental range.[18]

5.5.2.2. Combined mechanisms of boundary, hydrated and elastohydrodynamic lubrication

For a real gel, not only the interfacial interaction, but also the real contact area between two gels, should play an important role in the friction. Similarly to the situation found between two solids, the real contact area at the gel-gel or gel-solid interface is much smaller than the apparent contact area. This real contact is substantially dependent on the normal stress P, gel elasticity E, and sample thickness h. In this section, we present how these parameters influence the frictional behavior between two like-charged repulsive gels with finite roughness.

First, we start with the friction between flat, charged gels in ideal contact. According to the discussion in the previous section, the friction in this case originates from hydrated lubrication by the formation of the electric double layer at the charged interface. For the ring-shaped gel rotating on the disc-shaped gel, the viscous friction is

$$\sigma_{vis} \approx \frac{2\eta\nu}{3(h + 2\xi)} \tag{7}$$

Then, we consider the friction between rough, charged surfaces. In this case, we present a combined mechanism of boundary, hydrated and elastohydrodynamic lubrication, in which the external normal pressure, the sample rigidity and thickness all play important roles.

It is known that even for a transparent gel, intrinsic spatial fluctuation in polymer concentration exists due to inhomogeneous crosslinking, as revealed by various light-scattering and neutron-scattering studies. The heterogeneous structure of a gel is also revealed as a distribution in surface elastic modulus and roughness, as measured by surface-topography mapping of several kinds of neutral gels by means of AFM. Although there is no direct information on the surface microstructure of polyelectrolyte gels in pure water, from the neutron scattering results of the bulk gel that show a strong heterogeneity, there is no doubt that polyelectrolyte gels have a finite roughness and distribution of elastic modulus, similar to neutral gels. Here, we extend the electrical double layer (EDL) theory to charged surfaces with roughness.

When the nominal normal pressure $P \ll E$, the contact between gels with roughness is similar to that between solids, i.e., the asperities make contact first, and the load is sustained by the asperities. Several theories on the contact of nominally flat surfaces with random roughness have shown that (1) the real contact area A_s is much smaller than the nominal contact area A, and A_s is linearly proportional to the nominal pressure P; (2) the mean contact pressure P_{con} at the asperities is independent of P but depends on the material properties (the elastic modulus E and the Poisson ratio ν) and the topological feature of the surfaces.[53,57,58,59]

The friction mechanism consists of two components: one is from the asperities at contact, and the other is from the part between the asperities: $\sigma = \sigma_s \frac{A_s}{A} + \sigma_{vis}(1 - \frac{A_s}{A})$.

Then we can get[56]:

$$\sigma \approx \sigma_0 + \sigma_{vis} \quad (P \ll E) \tag{8}$$

Here, $\sigma_0 = \sigma_s(A_s/A)$. The first part may be due to boundary lubrication and be velocity-independent and the second is hydrated lubrication.[56] In

the next three sections, the effects of P, E and gel thickness are investigated
and discussed in terms of this combined mechanism.

5.5.2.3. *Effect of normal pressure*

Figure 5.17 shows the frictional behavior for a ring-shaped PNaAMPS gel
rotationally sliding against a disc-shaped PNaAMPS gel in water under
various normal pressures. Supposing the surfaces of the gels are ideally flat,
we can obtain the theoretical frictional stress from Eq. 7. The deviation
from the linear relationship between the frictional force and the sliding
velocity, especially at low velocities, apparently indicates that the friction
between the gels is not simply due to hydrated lubrication.

   If the gel has a finite surface roughness, the local pressure applied at the
asperities is proportional to E, irrespective of P. Accordingly, the combined
mechanism of *boundary* lubrication and *hydrated* lubrication may be applied
for rigid gels.

   As shown in Fig. 5.17, the similar P dependence of h_{exp} and $h + 2\xi$
indicates that the gel surfaces are not very rough, and that the contact
region of the asperities, and therefore the load sustained by the asperities,
is very small.

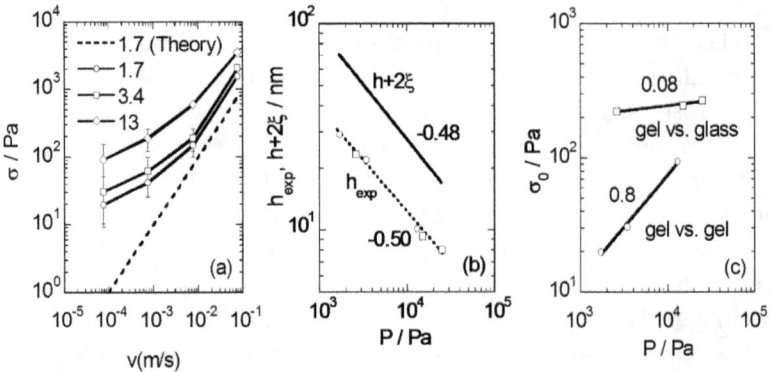

Fig. 5.17.   Frictional behaviors for a ring-shaped PNaAMPS gel rotationally sliding
against a disc-shaped PNaAMPS gel in water. (a) Velocity dependences of frictional
stress σ. Numbers in the figures denote the nominal pressure P in kPa; (b) P dependence
of hydrated lubricating layer thickness h_{exp}, where h_{exp} is estimated using the relation
$\sigma - \sigma_0 = 2\eta v/3h_{exp}$ for gel vs. gel (o) and for gel vs. glass ($\square$), the solid line is the
theoretical value $h + 2\xi$, where h is calculated by Eq. 4, and $2\xi = 3.2\,\text{nm}$ for q = 27.
(c) P dependence of the frictional stress σ_0, which is obtained from σ at the lowest sliding
velocity ($\nu = 7.5 \times 10^{-5}$ m/s) in (a). Swelling degree q: 27; sample thickness: 3.1 mm;
temperature: 25°C. Reproduced with permission from Ref. 56.

According to the contact theory for solids, local contact pressure P_{con} is insensitive to P. This suggests that the σ_s of boundary lubrication is not sensitive to P, while $\sigma_0 = \sigma_s A_s / A$ increases linearly with P. However, as shown in Figure 5.17c, the P dependence of the velocity-independent term σ_0 is quite different in the case of gel vs. gel friction ($\sigma_0 \propto P^{0.8}$) and gel vs. glass friction ($\sigma_0 \propto P^{0.08}$). Furthermore, the latter reveals a higher σ_0 value. This indicates that the contact between the asperities for gel-gel and for gel-glass could not be explained by a purely interfacial geometrical effect. Other effects, such as the resistance of counter-ions or the heterogeneity of the gel, may play roles that have yet to be proven.[56]

5.5.2.4. *Effect of gel modulus*

Figure 5.18 reveals the sliding velocity dependence of the dynamic frictional stress when the two PNaAMPS gels with various E are slid against each other under a constant nominal pressure. With a decrease in E (corresponding to an increase in q), the contribution from boundary lubrication becomes less important even at very low sliding velocities, due to the decrease in local contact pressures at asperities. A large discrepancy is observed between observation and theory in the low-velocity region. The discrepancy is reduced with decreasing E.[56]

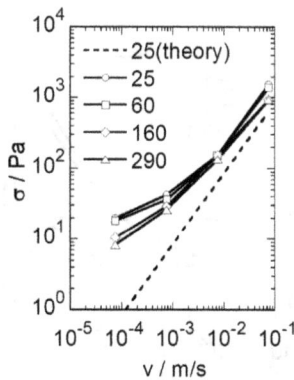

Fig. 5.18.   Frictional behavior of ring-shaped PNaAMPS gel of various E (or q) rotationally sliding against the disc-shaped PNaAMPS gel in water. Velocity dependences of frictional stress σ; numbers in the figure indicate swelling degree q; the dashed line indicates the theoretical data using $h + 2\xi = 74$ nm. Sample thickness: 3.1–3.5 mm, nominal pressure P: 1.7 kPa, temperature: 25°C. Reproduced with permission from Ref. 56.

5.5.2.5. *Effect of gel thickness*

When the gel thickness decreases, several macroscopic effects occur in contact. (1) misalignment of the two sliding surfaces; (2) unevenness in the sample thickness; (3) constrained deformation under compression. For (3), since one side of the gel surface is glued to the parallel plate, the homogeneous expansion of the gel in the lateral direction under compression is constrained.[60] All of these effects will influence the contact at the interface, and therefore, change the frictional behavior.[56]

As shown in Fig. 5.19, when the sample thickness d decreases, the friction behavior dramatically changes for a gel with $E = 640$ kPa. For a sample with thickness $d = 1$ mm, the frictional stress was not found to depend on the sliding velocity over the entire range of velocities investigated. It has been found that the thickness dependence of the friction mechanism varies as follows: (1) $d = 1 - 2$ mm, solid-like (boundary lubrication) friction over the entire range of velocities; (2) $d = 2 - 3$ mm, minimum friction at an intermediate velocity; (3) $d > 3$ mm, solid-liquid combined behavior as discussed in the previous part of the paper (Fig. 5.19).

A critical thickness is clearly indicated in Fig. 5.20, below which the frictional stress remains low due to boundary lubrication for rigid gels. This critical thickness for transition from solid-like to liquid-like friction mechanism is dependent on the softness of the gel: the softer the gel, the smaller is the critical thickness.

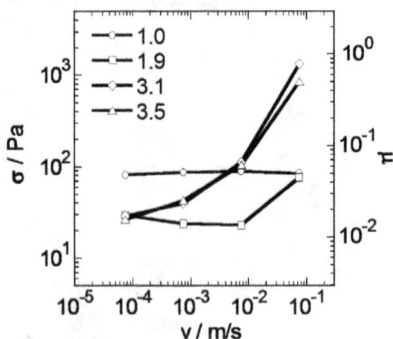

Fig. 5.19.   Velocity dependence of frictional stress σ and frictional coefficient μ for a ring-shaped PNaAMPS gel rotationally sliding against a disc-shaped PNaAMPS gel with different thickness in water. The numbers in the figures indicate the thickness of the gels in mm. Nominal pressure P: 1.7 kPa; swelling degree of gel: 27, temperature: 25°C. Reproduced with permission from Ref. 56.

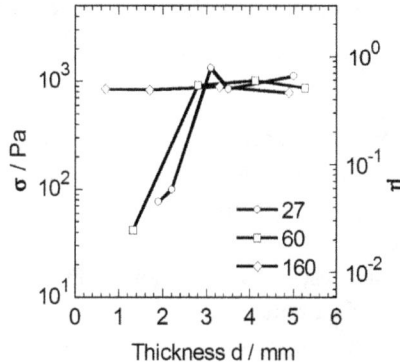

Fig. 5.20. PNaAMPS gel thickness dependences of frictional stress σ at the high-velocity limit of 0.1 m/s. Numbers in the figures indicate the swelling degree q of the PNaAMPS gels. Nominal pressure P: 1.7 kPa, temperature: 25°C. Reproduced with permission from Ref. 56.

All the studies confirm that the frictional stress between like-charged gels is composed of two components: a velocity-independent component as a result of boundary lubrication between highly compressed contacts of asperities, and a velocity-dependent component as a result of lubrication due to the liquid that exists interstitially at the repulsive interface. The mechanism of the formation of the lubricating layer h is different, depending on the softness and thickness of the gel. For a soft and thick gel, where macroscopic geometric effects, such as misalignment of the two surfaces, unevenness of sample thickness, and asymmetric deformation of the sample, are completely compensated by the deformation of the gel, h is determined by the balance between the local pressure and osmotic repulsion, which is independent of velocity. This type of lubrication is referred as *hydrated* lubrication, as it is specific to gels. For a rigid and thin gel, where macroscopic geometric effects play a dominant role in contact, h is velocity dependent due to the entrainment of the liquid from the outside, which is *elastohydrodynamic* lubrication. And it is similar as the *elastohydrodynamic* lubrication between elastic solids in a liquid.[56]

Many effects, such as heterogeneity of gel, counter-ions pushing back the gel surface and measurement conditions, will affect the frictional properties of gel. Since the boundary-lubrication component completely disappears for a gel with many dangling chains on its surface,[21] the intrinsic heterogeneous network structure of the gel may be the origin of the boundary lubrication for like-charged gels.

5.6. Friction of Hydrogels with Surface-Modified Structure

5.6.1. *Effects of template on hydrogel friction*

It has been discovered that the surface structure, and therefore friction, of a gel are strongly dependent on the substrate on which the gel is synthesized.[21] Such a substrate template effect is observed in a wide variety of hydrogels prepared from water-soluble vinyl monomers, such as the sodium salt of styrene sulfonate, acrylic acid, and acrylamide, on various hydrophobic substrates, such as Teflon, polyethylene, polypropylene, poly(vinyl chloride), and polymethyl methylacrylate (PMMA).[21]

Gels that are synthesized between a pair of glass substrates have a mirror-like smooth surface. This is the same when a hydrogel is synthesized on other hydrophilic substrates, such as mica and sapphire. However, even for the same chemical structure, a hydrogel exhibits an eel-like slimy surface when it is synthesized on hydrophobic substrates, such as Teflon and polystyrene (PS). The differences in the surface nature of the gels synthesized on different substrates are so clear that they can be easily distinguished by touching with one's finger. If a hydrogel is synthesized between two plates: one is hydrophobic and the other is hydrophilic, heterogeneous gelation occurs.[61] After equilibrated swelling in water, the gel exhibits a significant curvature as shown in Fig. 5.21, where the gel surface formed on the Teflon surface is always the outside of the curvature and that on the glass is the inside. It is because the gel surface close to the Teflon has a higher swelling while the one next has a lower value. It has been clarified that the gel has a gradient structure and the surface formed on Teflon (or other hydrophobic substrates) has a low cross-linking density, with branched dangling polymer chains.[62]

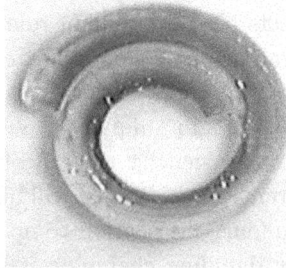

Fig. 5.21.   Photograph of a PAMPS gel prepared between a Teflon plate and a glass plate after swelling in water. Reproduced with permission from Ref. 61.

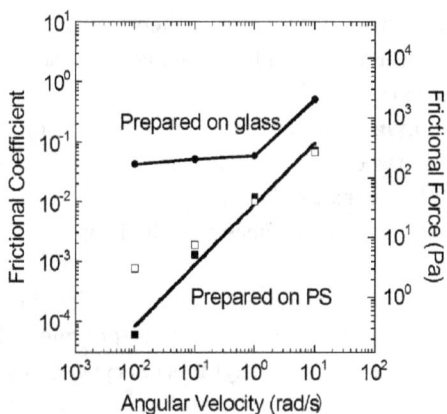

Fig. 5.22. Angular-velocity dependencies of the frictional coefficient of PAMPS gels slid against a glass plate in water under a normal pressure of 4×10^3 Pa measured by a rheometer. (●) prepared on glass, swelling degree, 21; (■) prepared on PS, swelling degree, 27; and (□) containing linear polymer chains, Sample size, 1 cm × 1 cm, swelling degree, 15. Reproduced from Ref. 21.

The frictional force and frictional coefficient of PAMPS gels synthesized on a glass plate and on a PS plate were measured against a glass plate in water by a rheometer (Fig. 5.22).[21] The frictional stress of the gel prepared on a PS substrate shows a lower value than that prepared on a glass substrate. Especially in the low-velocity range, the frictional stress of the gel prepared on the PS substrate reaches a value as low as 1Pa, a value equivalent to the shear stress on the wall of blood vessels.[13] The frictional coefficient of the gel prepared on PS reaches 10^{-4}, at least two orders lower in magnitude than that of the gel synthesized on glass at the low velocity range.

The reduction in friction is attributed to the presence of branched dangling chains on the gel surface prepared on the hydrophobic substrate, as revealed by the result for the PAMPS gels containing free linear PAMPS polymer chains prepared on the glass plate, which showed similarly low friction coefficients (Fig. 5.22).

5.6.2. *Hydrogel surface with polyelectrolyte brushes*

It is found that the dramatic reduction of friction, as shown in Fig. 5.22, is due to the presence of dangling surface chains. Needless to say, this kind of friction-reduction effect only occurs when the polymer chains are non-adhesive to the countersurface. The presence of adhesive dangling chains, in contrast, increases the friction.

The effect of polymer brushes on the reduction of sliding friction has also been observed between solid surfaces by surface forces apparatus (SFA) measurements.[63–65] For example, Raviv *et al.* reported a massive friction reduction between mica surfaces modified by repulsive polyelectrolyte brushes in water.[65]

To quantitatively investigate the relationships between the friction and the polymer brush properties, such as polymer chain length, density, hydrogels of poly(2-hydroxyethyl methacrylate) (PHEMA) with well-defined polyelectrolyte brushes of poly(sodium 4-styrenesulfonate) (PNaSS) of various lengths (molecular weights) were synthesized, keeping the distance between the polymer brushes constant at *ca* 20 nm.[27] The effect of polyelectrolyte brush length on the sliding friction against a glass plate, an electrostatic repulsive solid countersurface, was investigated in water, as shown in Fig. 5.23. It is found that the presence of a polymer brush can dramatically reduce the friction when the polymer brushes are not very long. With an increase in the length of the polymer brush, this drag reduction effect only works at a low sliding velocity, and the gel with long polymer brushes even shows a higher friction than that of a normal network

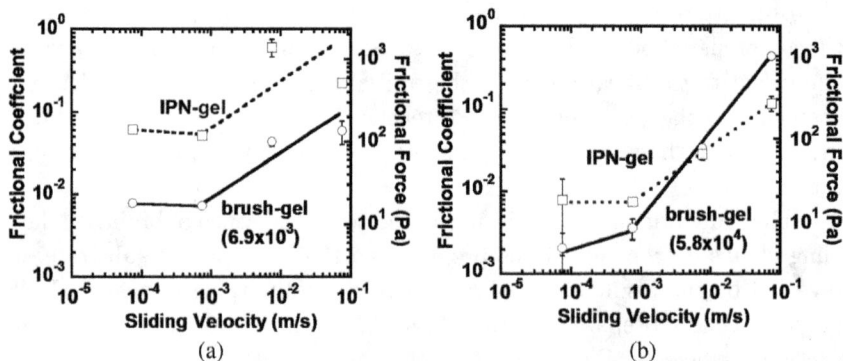

Fig. 5.23.   Velocity dependences of the frictional coefficient for the gels sliding on a glass surface in pure water at 25°C under a normal pressure of 2.3 kPa. (○) Brush-gels consist of neutral polymer PHEMA as network and negatively charged PNaSS as brushes; (□) IPN-gels consist of PHEMA and PNaSS with interpenetrated network structure. Numbers in the figure are molecular weights of polymer brushes. (a) R = 1.2%. Brush-gel, q = 10.5. IPN-gel, q = 11. (b) R = 15%. Brush-gel, q = 44. IPN-gel, q = 52. R is NaSS/HEMA molar ratio and q is swelling degree of gels. Reproduced with permission from Ref. 27.

gel at a high sliding velocity. The strong polymer length and sliding velocity dependence indicate a dynamic mechanism of the polymer brush effect.

Supposing the hydrated water layer behaves as lubricant, the apparent lubricating layer thickness L_T can be estimated from the frictional force σ by $\sigma = \eta v / L_T$. Here, η is the solvent viscosity and v the sliding velocity. Figure 5.24 ($\circ$) shows that the lubricating layer thickness L_T thus obtained monotonically increases with the sliding velocity from ca 5 nm to ca 200 nm for a brush-gel with a length of 8.4 nm, indicating a shear-thinning effect. For the brush-gel with a brush length of 70.2 nm, longer than the next neighboring distance (20 nm), L_T shows a maximum with the sliding velocity, while the IPN-gel with a similar composition and swelling degree q shows a monotonic increase in L_T with v (Fig. 5.24), ($\bullet$). Figure 5.24 also shows that for the same PNaSS content, the brush-gel shows a greater L_T than that of the IPN-gel when the brush is short (8.4 nm) or when the sliding velocity is not very high for a long brush (70.2 nm), indicating the drag-reduction effect of the brush. What is interesting is that the change in L_T with sliding velocity is about 100 nm, much larger than the brush length (Fig. 5.24(a)) or the mesh size of the network (Fig. 5.24(b)).

Two reasons might be responsible for the friction reduction: One is associated with the enhanced thickness of the solvent layer between branched

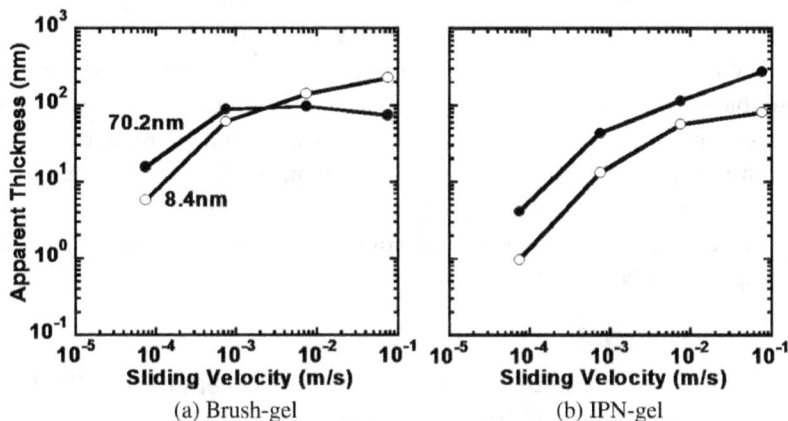

Fig. 5.24. Sliding velocity dependence of the apparent hydrated layer thickness for brush-gels (a) and IPN-gels (b) of similar degree of swelling q or NaSS/HEMA molar ratio R. Numbers in figure are contour lengths of PNaSS brushes. ($\circ$): q = 19. Brush-gel, R = 5.9%; IPN-gel, R = 13%. ($\bullet$): R = 15%. Brush-gel, q = 44; IPN-gel, q = 52. Reproduced with permission from Ref. 27.

dangling polymer chains and the sliding countersurface. When the interfacial interaction is repulsive, a water layer is retained at the interface, even under a large normal load, yielding low friction. At the same pressure, the static solvent layer thickness should be the same for both a chemically cross-linked gel and a gel having branched dangling polymer chains on its surface. However, during sliding, the polymer brushes are deformed more easily than the cross-linked network, and this would increase the effective thickness of the lubricating layer and reduce the shear resistance. Another reason is associated with the decreased coherence length of the surface layer of the gel. A branched surface should have a shorter coherence length than a cross-linked network surface due to the freedom of the one-end-free polymer chains. This ensures very homogeneous contact pressure and favors the EDL formation over the whole interface and eliminates the boundary lubrication component that observed between normal polyelectrolyte gels.

5.7. Application of Robust Gels with Low Friction as Substitutes for Biological Tissues

Most biological tissues are in a gel state, for example, articular cartilage is a natural fiber-reinforced hydrogel composed of proteoglycans, type II collagen, and approximately 70% water. The normal cartilage tissue contributes significantly to the joint functions involving ultra-low friction, distribution of loads, and absorption of impact energy. When normal cartilage tissues are damaged, it is extremely difficult to regenerate these tissues with currently available therapeutic treatments. Therefore, it is important to develop substitutes for the normal cartilage tissue as a potential therapeutic option. The important characteristic of materials required for load-bearing tissue replacements is their comparable mechanical property with the native tissue. For example, the severe loading condition imposed on human articular cartilage is as high as 1.9–14.4 MPa.[5]

5.7.1. Robust gels with low friction — an excellent candidate material as artificial cartilage

It has been described that a hydrogel with branched dangling polymer chains on its surface can effectively reduce the surface sliding frictional coefficient to a value as low as 10^{-4}.[21] Unfortunately, conventional hydrogels are mechanically too weak to be practically used in any stress or strain-bearing applications, which hinders the extensive application of hydrogels as industrial and biomedical materials. Design and production of hydrogels with a low surface friction and high mechanical strength are crucially important in the biomedical applications of hydrogels as contact lens, catheter, artificial articular cartilage, and artificial esophagus.[66]

Gong and co-workers have reported a novel method to obtain hydrogels with extremely high mechanical toughness.[67,68] The hydrogel is composed of two kinds of hydrophilic polymers (double-network (DN) structure) which, despite containing 90% water, have 0.4–0.9 MPa elastic modulus and exhibit the compressive fracture strength as high as several tens of MPa. The point of this discovery is that a DN is composed of a combination of a stiff, brittle first network with a soft, ductile second network. The molar composition and the cross-linking ratio of the first and second networks are also important. A DN hydrogel is completely different in concept from the common interpenetrating polymer networks or a fiber-reinforced hydrogel, which is only a linear combination of two component networks.

Based on these researches, the new soft- and wet- materials, with both lower friction and high strength are synthesized by introducing a weakly cross-linked PAMPS network (to form a triple-network, or TN, gel) or a non-cross-linked linear polymer chain (to form a DN-L gel) as a third component into the optimally tough PAMPS/PAAm DN gel.[69] The TN and DN-L gels were synthesized by UV irradiation after immersing the DN gels in a large amount of a third solution of 1M AMPS and 0.1 mol% 2-oxoglutaric acid with (TN) and without the presence of 0.1 mol% MBAA (DN-L). The mechanical properties of the gels are summarized in Table 5.1. After introducing cross-linked or linear PAMPS to DN gel, the fracture strength of TN and DN-L gels remains on the order of MP, and elasticity of gels are higher than that of DN gel ($\sim$2 MPa). In addition, the fracture strength of DN-L remarkably increases because PAMPS linear chains can effectively dissipate the fracture energy.

Table 5.1. Mechanical properties of DN, TN, and DN-L gels. Reproduced with permission from Ref. 29.

Gel	Water Content [wt.%]	Elasticity [MPa]	Fracture stress σ_{max} [MPa]	Fracture Strain λ_{max} [%]
DN	84.8	0.84	4.6	65
TN	82.5	2.0	4.8	57
DN-L	84.8	2.1	9.2	70

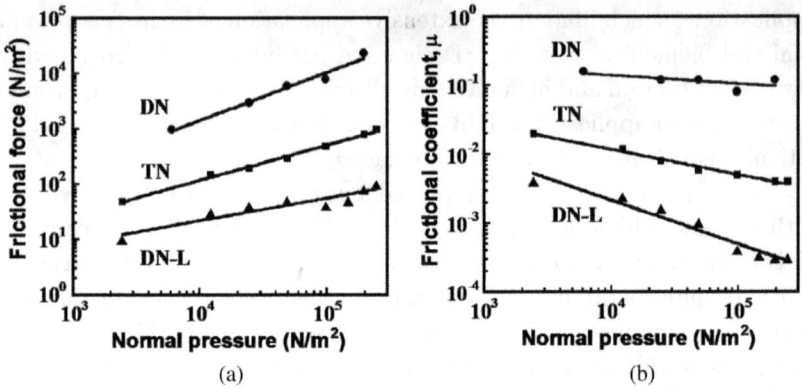

Fig. 5.25.   Normal pressure dependences of frictional force (a) and frictional coefficient (b) of hydrogels against a glass plate in pure water. Sliding velocity: 1.7×10^{-3} [m/s]. Symbols denote DN (•), TN (■), and DN-L(▲) gels, respectively. Reproduced with permission from Ref. 29.

Figures 5.25(a) and (b) show the frictional forces (F) and frictional coefficients (μ) of the three kinds of gels as a function of normal pressure (P). The gels were slid against a glass plate in water. It clearly shows that the frictional coefficient decreases in the order DN > TN > DN-L, indicating that introducing PAMPS, especially linear PAMPS, as the third network component clearly reduces the friction coefficient of the gels.

The DN gel has a relatively large value of friction coefficient ($\sim 10^{-1}$) since the second network, non-ionic PAAm, dominates the surface of the DN gel, which is adsorptive to the glass countersurface.

However, when the PAMPS network is introduced into the DN gel as a third component, the friction coefficient of the TN gel decreases to $\sim 10^{-2}$, which is two orders of magnitude lower than that of the DN gel, since the surface of the TN gel is dominated by PAMPS, resulting in a repulsive interaction with the glass countersurface and reducing the frictional force. Furthermore, when linear PAMPS chains are introduced into the surface of a DN gel, the friction coefficient reduces significantly to $\sim 10^{-4}$, which is one to three orders of magnitude less than that of TN and DN gels. This demonstrates that the linear PAMPS chains on the gel surface reduce the friction force due to further repulsive interactions with the glass countersurface.[21]

It should be emphasized that the lower friction coefficient of the DN-L gel can be observed under a pressure range of $10^3 - 10^5$ Pa. The results demonstrate that the linear polyelectrolyte chains are still effective in retaining lubrication even under an extremely high normal pressure.

5.7.2. *Wear properties of robust DN gels*

For application of DN gels as artificial articular cartilage, it is critical to evaluate wear properties, because articular joints are subject to rapid changes in shear-force magnitude for millions of cycles over a lifetime. However, there are no established methods for evaluating the wear properties of a gel. Pin-on-flat wear testing, which has been used to evaluate the wear of ultra-high molecular weight polyethylene (UHMWPE), which is the only one established rigid and hard biomaterial used in the artificial joint, has been used to evaluate the wear properties of DN gels.

The wear properties of four kinds of DN gels, which were composed of synthetic or natural polymers, were evaluated.[68] The first gel was PAMPS/PAAm DN gel, which consists of poly (2-acrylamide-2-methyl-propane sulfonic acid) and polyacrylamide. The second gel was PAMPS/PDMAAm DN gel, which consists of poly (2-acrylamide-2-methyl-propane sulfonic acid) and poly (N,N'-dimethyl acrylamide). The third gel was cellulose/PDMAAm DN gel, which consists of bacterial cellulose and poly-dimethyl-acrylamide. The fourth gel was cellulose/gelatin DN gel, which consists of bacterial cellulose and gelatin. These 4 unique gel materials have great potential for application as an artificial cartilage.

Over one million friction test cycles, which was equivalent to 50 km friction (50 mm $\times$ 10^6), the maximum wear depths of the PAMPS/PAAm, PAMPS/PDMAAm, BC/PDMAAm, and BC/gelatin gel were 9.5, 3.2, 7.8, and 1302.4 μm, respectively. It is impressive that the maximum wear depth of PAMPS/PDMAAm DN gel is similar to the value for UHMWPE (3.33 μm). In addition, although the maximum wear depths of the PAMPS/PAAm DN gel and BC/PDMAAm DN gel were about 2–3 times higher than that of UHMWPE, these gels could bear one million friction cycles. The results demonstrate that PAMPS/PAAm, PAMPS/PDMAAm and BC/PDMAAm DN gels are resistant to wear to a greater degree than conventional hydrogels, and PAMPS/PDMAAm DN gel can potentially be used as replacement material for artifical cartilage. On the other hand, BC/gelatin DN gel, which is composed of natural materials, shows extremely poor wear properties in comparison to other DN gels. The poorer wear properties of the BC/gelatin DN gel can be ascribed to reasons such as its relatively low water content, higher friction coefficient, and the fact that it is easily roughened by abrasion.

In order to design materials that can be potentially used as artificial cartilage, appropriate viscoelasticity, high mechanical strength, durability to repetitive stress, low friction, high resistance to wear, and resistance to biodegradation within the living body are required. It is difficult to develop a gel material that satisfies even two of these requirements at the same time. However, recent breakthroughs in synthesizing mechanically strong hydrogels changes the conventional gel concept, and opens a new era of soft and wet materials as substitutes for articular cartilage and other tissues. The recent research shows that the friction coefficient of the artificial to normal cartilage articulation (0.029) was significantly lower than that of the normal to normal cartilage articulation (0.188), which proves that with excellent biocompatibility, resistance to biodegradation frictional wear and excellent frictional properties, the PAMPS/ PDMAAm DN gel is an attractive material to develop artificial cartilage in the future.[71]

5.8. Summary

The frictional behavior of polymer gels is more complex than that of solids, and is dependent on the chemical structure and properties of the gels, geometry of gels, the surface properties of sliding countersurfaces, and the measurement conditions. The surface friction of gels against a smooth countersurface can be divided into two categories, i.e., adhesive gels and repulsive gels, determined by the combination of the gels and the opposing surfaces. For adhesive gels, friction originates from two contributions: namely surface adhesion and hydrated lubrication. The former is dominant at low sliding velocities and the latter is dominant at high velocities. The friction shows an S-shape curve with the sliding velocity. A transition in friction occurs at the sliding velocity characterized not only by the mesh size and polymer chain relaxation time of the gel but also by the geometry and substrate roughness of the gel. For a repulsive gel, the friction is believed to be due to lubrication by the hydrated water layer. The friction shows a monotonic increase with the sliding velocity due to its hydrodynamic nature. The presence of repulsive dangling chains on gel surfaces dramatically reduces the friction and the brush-gels show a friction coefficient as low as 10^{-4}, which is comparable to that found for animal joints.

Looking for materials with low friction has been one of the classical and perpetual research topics for materials scientists and engineers. Despite many efforts, it has been shown that surface modification or adding lubricants are only limited approaches to the reduction of friction to low values, which normally show a frictional coefficient $\mu \sim 10^{-1}$ even in the presence of a lubricant. The recent discovery of an extremely low-friction gel should now enable hydrogels to find wide application in many fields where low friction is required.

References

1. D. Dowson, *History of Tribology*, Longman, London (1979).
2. N. P. Suh, *Tribophysics*, Prentice-Hall, Englewood Cliffs, NJ (1986).
3. A. Dhinojwala, L. Cai and S. Granick, *Langmuir* 12, 4537 (1996).
4. C. W. McCutchen, *Wear* 5, 1 (1962).
5. C. W. McCutchen, *Lubrication of Joints, the Joints and Synovial Fluid*, Academic Press, New York (1978).
6. D. Dowson, A. Unsworth and V. Wright, *J Mech Eng Sci* 12, 364 (1970).
7. G. A. Ateshian, H.Q. Wang and W. M. Lai, *J Tribol* 120, 241 (1998).
8. W. A. Hodge, R. S. Fijian, K. L. Carlson, R. G. Burgess, W. H. Harris and R. W. Mann, *Proc Natl Acad Sci USA* 83, 2879 (1986).
9. A. J. Grodzinsky, *CRC Crit Rev Biomed Eng* 9, 133 (1985).
10. M. D. Buschmann and A. J. Grodzinsky, *J Biomech Eng* 117, 179 (1995).
11. E. M. Wojtys and D. B. Chan, *Instr Course Lect* 54, 323 (2005).
12. B. N. J. Persson, *Sliding Friction: Physical Principles and Applications*, 2nd ed., NanoScience and Technology Series, Springer, Berlin (1998).
13. Y. C. Fung, *Biomechanics: Mechanical Properties of Living Tissues*, Springer-Verlag New York, Inc. 2nd Edition (1993).
14. J. P. Gong, Y. Katsuyama, T. Kurokawa and Y. Osada, *Adv Mat* 15(14), 1155–1158 (2003).
15. J. P. Gong, *Soft Matter* 2, 544–552 (2006).
16. J. P. Gong, M. Higa, Y. Iwasaki, Y. Katsuyama and Y. Osada, *J Phys Chem B* 101, 5487 (1997).
17. J. P. Gong and Y. Osada, *J Chem Phys* 109, 8062 (1998).
18. J. P. Gong, G. Kagata and Y. Osada, *J Phys Chem B* 103, 6007 (1999).
19. J. P. Gong, Y. Iwasaki, Y. Osada, K. Kurihara and Y. Hamai, *J Phys Chem B* 103, 6001 (1999).
20. J. P. Gong, Y. Iwasaki and Y. Osada, *J Phys Chem B* 104, 3423 (2000).
21. J. P. Gong, T. Kurokawa, T. Narita, K. Kagata, Y. Osada, G. Nishimura and M. Kinjo, *J Am Chem Soc* 123, 5582 (2001).
22. G. Kagata, J. P. Gong and Y. Osada, *J Phys Chem B* 106, 4596 (2002).
23. G. Kagata, J. P. Gong and Y. Osada, *J Phys Chem B* 107, 10221 (2003).
24. T. Kurokawa, J. P. Gong and Y. Osada, *Macromolecules* 35, 8161 (2002).
25. T. Baumberger, C. Caroli and O. Ronsin, *Phys Rev Lett* 88, 75509 (2002).
26. T. Baumberger, C. Caroli and O. Ronsin, *Eur Phys J E* 11, 85 (2003).
27. Y. Ohsedo, R. Takashina, J. P. Gong and Y. Osada, *Langmuir* 20, 6549 (2004).
28. T. Tada, D. Kaneko, J. P. Gong, T. Kaneko and Y. Osada, *Tribol Lett* 17, 505 (2004).
29. D. Kaneko, T. Tada, T. Kurokawa, J. P. Gong and Y. Osada, *Adv Mater* 17, 535 (2005).

30. T. Kurokawa, T. Tominaga, Y. Katsuyama, R. Kuwabara, H. Furukawa, Y. Osada and J. P. Gong, *Langmuir* 21, 8643 (2005).

31. Y. Nitta, H. Haga and K. Kawabata, *J Phys IV* 12, 319 (2002).

32. Z. T. Jiang, T. Tominaga, K. Kamata, Y. Osada and J. P. Gong, *Coll Surf A: Physicochem Eng Aspects* 284, 56 (2006).

33. T. Tominaga, N. Takedomi, H. Biederman, H. Furukawa, Y. Osoda and J. P. Gong, *Soft Matter* 4, 1033–1040 (2008).

34. T. Tominaga, T. Kurokawa, H. Furukawa, Y. Osoda and J. P. Gong, *Soft Matter* 4, 1645–1652 (2008).

35. P. G. de Gennes, *Scaling Concept in Polymer Physics*, Cornell University Press: Ithaca, NY (1979).

36. A. Shallamach, *Wear* 6, 375 (1963).

37. Y. B. Chernyak and A. I. Leonov, *Wear* 108, 105 (1986).

38. A. R. Savkoor, *Wear* 8, 222 (1965).

39. K. C. Ludema and D. Tabor, *Wear* 9, 329 (1966).

40. K. Vorvolakos and M. K. Chaudhury, *Langmiur* 19, 6778 (2003).

41. A. Martin, J. Clain, A. Buguin and F. Brochard-Wyart, *Phys Rev E* 65, 031605 (2002).

42. P. G. de Gennes, F. Brochard-Wyart and D. Quéré, *Capillarity and Wetting Phenomena: Drops, Bubbles, Pearls, Waves*, Springer, New York (2003).

43. F. Heslot, T. Baumberger, B. Perrin, B. Caroli and C. Caroli, *Phys Rev E*, 49, 4973 (1994).

44. E. Ruckenstein and P. Rajora, *J Coll Inter Sci* 96, 488 (1983).

45. D. Tretheway and C. Meinhart, *Phys Fluids* 14, L9 (2002).

46. A. Steinberger, C. Cottin-Bizonne, P. Kleimann and E. Charlaix, *Nat Mater* 6, 665 (2007).

47. K. Watanabe, Y. Udagawa and H. Udagawa, *J Fluid Mech* 381, 225 (1995).

48. J. Ou, B. Perot and J. P. Routhstein, *Phys Fluids* 16, 4635 (2004).

49. C.-H. Choi and C.-J. Kim, *Phys Rev Lett* 96, 066001 (2006).

50. N. Ishida, T. Inoue, M. Miyahara and K. Higashitani, *Langmuir* 16, 6377 (2000).

51. J. W. G. Tyrrell and P. Attard, *Langmuir* 18, 160 (2002).

52. N. Mishchuk, J. Ralston and D. Fornasiero, *J Phys Chem A* 106, 689 (2002).

53. B. N. J. Persson, *J Chem Phys* 115, 3840 (2001).

54. M. Du, Y. Maki, T. Tominaga, H. Furukawa, J. P. Gong, Y. Osada and Q. Zheng, *Macromolecules* 40(12), 4313–4321 (2007).

55. J. P. Gong, G. Kagata, Y. Iwasaki and Y. Osada, *Wear* 251, 1183–1187 (2001).

56. S. Oogaki, G. Kagata, T. Kurokawa, S. Kuroda, Y. Osada and J. P. Gong, *Soft Matter* 5, 1879–1887 (2009).

57. K. L. Johnson, *Contact Mechanics*, Cambridge: Cambridge University Press (1966).

58. J. A. Greenwood and J. B. P. Williamson, *Proc R Soc* A 295, 300 (1966).

59. B. N. J. Persson, *Phys: Condens Matter* 17, R1 (2005).

60. A. N. Gent and P. B. Lindley, *Proc Instn Mech Engrs*, London, 173, 111 (1959).

61. A. Kii, J. Xu, J. P. Gong, Y. Osada and X. M. Zhang, *J Phys Chem B* 105, 4565 (2001).
62. T. Narita, A. Knaebel, J. P. Munch, S. J. Candau, J. P. Gong and Y. Osada, *Macromolecules* 34, 5725 (2001).
63. J. Klein, E. Kumacheva, D. Mahalu, D. Perahia and L. Fetters, *Nature* 370, 634 (1994).
64. G. S. Grest, *Adv Polym Sci* 138, 149 (1999).
65. U. Raviv, S. Giasson, N. Kampf, J. F. Gohy, R. Jerome and J. Klein, *Nature* 425, 163 (2003).
66. M. E. Freeman, M. J. Furey, B. J. Love and J. M. Hampton, *Wear* 241, 129 (2000).
67. J. P. Gong, Y. Katsuyama, T. Kurokawa and Y. Osada, *Adv Mater* 15, 1155 (2003).
68. Y.-H. Na, T. Kurokawa, Y. Katsuyama, H. Tsukeshiba, J. P. Gong, Y. Osada, S. Okabe and M. Shibayama, *Macromolecules* 37, 5370 (2004).
69. D. Kaneko, T. Tada, T. Kurokawa, J. P. Gong and Y. Osada, *Adv Mater* 17, 535 (2005).
70. H. Trieu and S. Qutubuddin, *Polymer* 36, 2531 (1995).
71. K. Arakaki, N. Kitamura, H. Fujiki, T. Kurokawa, M. Iwanoto, M. Ueno, F. Kanaya, Y. Osada, J. P. Gong and K. Yasuda, 2009 Wiley Periodicals, Inc. *J Biomed Mater Res* 93A, 110–1168 (2010).

Chapter 6

Aqueous Lubrication with Polymer Brushes

Suzanne Giasson
Université de Montréal, Département de Chimie
et Faculté de Pharmacie, C.P. 6128, succursale Centre-ville
Montréal (Québec) H3C 3J7, Canada
suzanne.giasson@umontreal.ca

Nicholas D. Spencer
Laboratory for Surface Science and Technology
Department of Materials, ETH Zurich
Wolfgang-Pauli-Strasse 10, CH-8093 Zurich, Switzerland
nspencer@ethz.ch

6.1. Introduction

6.1.1. *Polymer brushes*

When designing artificial lubricants for an aqueous environment, Nature is the obvious model. Nature's highly effective lubrication mechanisms are mostly based on sugar-decorated polymers — glycoproteins – that utilize the facility with which sugars can be hydrated, in order to hold a significant amount of water at the surface to be lubricated.[1] Charge, hydrophobicity, and hydrogen bonding all play significant roles in the behavior of natural lubricant molecules, although the precise way in which they function is still debated (see Chaps. 1 and 2). Nevertheless, in an attempt to mimic the lubricating function of the natural molecules, humans have resorted to a class of molecules known as polymer brushes, which are generally simpler, certainly easier to synthesize, and whose behavior is somewhat better understood than that of their natural counterparts.

When polymer chains are attached to a surface in the presence of a "good" solvent, and are spaced within a distance, s, approximately less

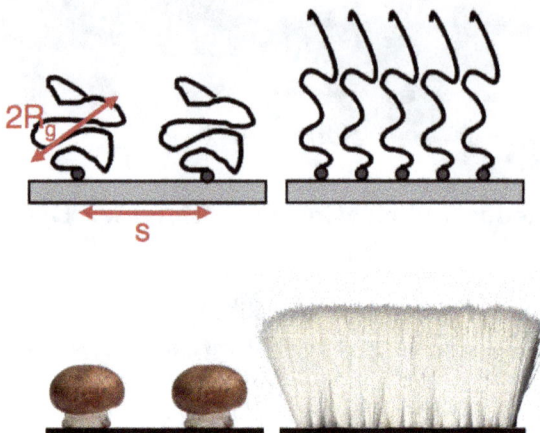

Fig. 6.1.   End-grafted chains of polymers in the presence of a good solvent. When $s >$ $2R_g$, the chains' conformation resembles that of the free molecules in solution, with radius of gyration, R_g. This is designated as the "mushroom" conformation. If $s < 2R_g$, the chains extend into the solvent, forming a polymer "brush". Reproduced from Ref. 3 with kind permission.

than twice their radii of gyration,[2] R_g, determined when freely moving in a good solvent, they stretch out away from the surface, working against the entropy elasticity of the chains. The incentive for the formation of this "brush-like" configuration (Fig. 6.1) is the high energetic cost to the chains of interacting closely with each other.

For the widely spaced mushroom conformation, the maximum height of the polymer brush is independent of the number of attached chains per unit area, or grafting density. For neutral systems, when in the brush conformation, the height scales as the cube root of the grafting density,[4] σ, which is equivalent to $1/s^2$ (Fig. 6.2).[5]

For polymer brushes formed from polyelectrolyte chains, some extra considerations need to be taken into account, and theoretical approaches using self-consistent mean-field theory (SCFT), scaling laws, strong-stretching theory (SST), and molecular dynamics have been used to investigate the monomer distribution, or the density profile, throughout charged brushes.[6–10]

The counterion distribution can remain within the brush layer and make the brush globally neutral. However, the brush is swollen by the osmotic pressure of the counterions and is referred to as an osmotic brush. This is usually the case for highly charged brushes where simple scaling laws developed by Pincus predict a brush thickness that is independent of

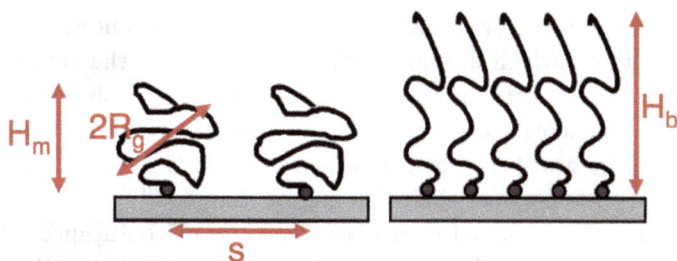

Fig. 6.2. Dependence of polymer height on chain spacing in both mushroom and brush conformations. The quantity $s/2R_g$ provides an indication of the extent of brush formation. In the mushroom conformation, the height of the polymer layer, H_m is independent of the grafting density, σ, i.e. $H_m \sim \sigma^0$, while for a neutral brush, the height, H_b is dependent on the cube root of the grafting density, i.e. $H_b \sim \sigma^{1/3}$. Reproduced from Ref. 3 with kind permission.

the grafting density[9]:

$$H \propto a N \alpha^{1/2} \tag{1}$$

where α is the degree of dissociation of the polymer chain (or fraction of monomers that are dissociated), N the degree of polymerization and a the monomer dimension. The Pincus model assumes that the monomer distribution is uniform through the brush. Equation 1 results from a balance between the swelling effect of the counterion entropy and the chain elasticity. On the other hand, when the counterion distribution extends beyond the brush, some of the free counterions are outside the brush, leaving a net charge at the outer brush surface. In this condition, referred to as the Pincus regime, and for weakly charged brushes, assuming that the monomer distribution is uniform through the brush, the brush height predicted by Pincus[9] is:

$$H \propto a^2 N^3 \sigma \alpha^2 \tag{2}$$

Viewed from the surrounding medium, the brush resembles a simple charged surface and long-ranged double-layer repulsion interactions arise for separation distances larger than twice the thickness of the brushes. At smaller separations, when two opposing brush layers are in physical contact, the interaction forces result from a balance between the osmotic, elastic and electrostatic contributions. According to these approaches and under salt-free conditions, the charged brush thickness H is expected to vary linearly with the degree of polymerization N when all polyelectrolyte counterions are located inside the brush, and to vary as $N^3 \sigma$ when some of these counterions extend outside the brush. In reality, however, the concentration profile throughout the brush exhibits a parabolic profile and there are thermal fluctuations, which are not taken into account in the simple scaling theories. A theoretical study using SCFT and SST has shown that thermal fluctuations create a small tail in the concentration profile extending beyond the classical brush height, which completely wipe out the Pincus regime unless the polymer chains are highly charged.[10]

The presence of added electrolytes and the pH both affect the electrostatic interactions within and between the polymer chains, the osmotic pressure of the counterions, and the degree of dissociation of the chains. Theoretical studies of polyelectrolyte brushes predict complex variation in brush height depending on the degree of dissociation of the chain (α), grafting density (σ), and the concentration of added salt (C_s).[8,9,11–14]

Mean-field theory has been used to predict the dependences of the brush height and degree of chain ionization on grafting density and ionic strength of the solution.[13] For annealed PE brushes at high grafting density and low salt concentrations, the brush thickness is expected to depend on the grafting density, pH and the salt concentration according to[13]:

$$H \propto N\sigma^{-1/3} \left(\frac{\alpha_s}{1-\alpha_s} C_{H^+} + C_s \right)^{1/3} \tag{3}$$

where α_s is the degree of dissociation of the isolated chains in solution. This regime is generally referred to as the annealed osmotic regime. For salt concentrations larger than the concentration of the brush counterions, electrostatic screening effects are expected to reduce the brush thickness according to[13]:

$$H \propto N\sigma^{1/3}\alpha^{2/3}C_s^{-1/3} \tag{4}$$

This regime is called the salted-brush regime.

For highly charged polyelectrolytes and when the added salt concentration (C_s) dominates over the counterions associated with polyelectrolyte layers, scaling laws developed by Pincus predict the brush height to vary with the salt concentration according to[9]:

$$H \propto \sigma^{1/3}NC_s^{-1/3} \tag{5}$$

6.1.2. *Brush preparation and characterization*

There are many ways to prepare polymer brushes and these can be generally grouped as "grafting-from" or "grafting-to"[15] (Fig. 6.3). Grafting-from

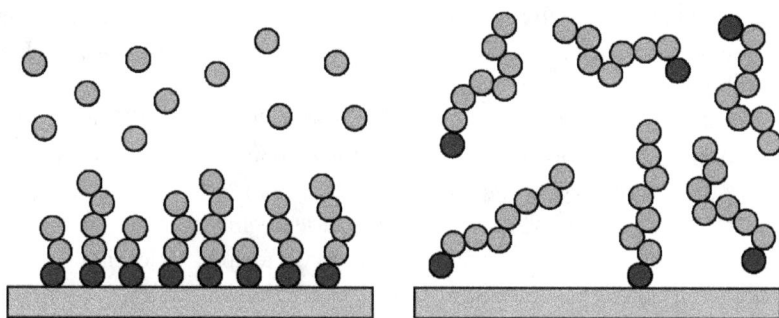

Fig. 6.3. "Grafting-from" (left) and "grafting-to" (right) approaches for the synthesis of polymer brushes. Reproduced from Ref. 3 with kind permission.

involves an initiator being immobilized on a substrate, from which a polymerization reaction takes place, the polymer actually growing out from the surface. Many different polymerization approaches can be adapted to grafting-from reactions, among them atom-transfer radical polymerization (ATRP)[16] and reversible addition-fragmentation chain transfer (RAFT).[17] Several strategies can be used to control the initiator grafting density on surfaces, including dilution of self-assembled monolayers of initiators with inert molecules, or Langmuir-Blodgett deposition.[18-21] Grafting-from has the advantage of producing a high attached-polymer-chain density (high σ), but clearly brush formation in this case requires a polymerization reaction to be carried out.

Grafting-to involves the pre-synthesis of end-functionalized polymers and their subsequent attachment to a surface. Examples would be the anchoring of thiol-modified poly(ethylene glycol) (PEG-thiol) molecules to a gold surface[22] or the anchoring of chlorodimethylsilyl end-functionalized polystyrene to a plasma-activated mica surface.[23] Clearly this is convenient, since the polymer synthesis can be performed once and brushes prepared many times. High values of σ are difficult to reach, since adsorbing chains steadily inhibit the adsorption of further chains due to steric-hindrance effects. By grafting polymer chains onto a backbone, forming a bottle-brush-like molecule, the situation in grafting-to can be improved, since by adsorption of the backbone, many chains can be adsorbed at once.

The brush conformation can also be obtained by means of diblock copolymers using the "grafting-to" approach. The brushes are formed via a self-assembly process at the surface, similar to that of a surfactant, when using the appropriate size and solvent selectivity for each polymer block.

In aqueous media, hydrophobic-b-hydrophilic copolymers, which behave as a macrosurfactant, are often used to form polymer brushes, with the hydrophobic block anchored on hydrophobic surfaces and the hydrophilic block protruding into the aqueous medium.[24] Such amphiphilic copolymers can also be grafted to a surface in a homogeneous monolayer with a controlled surface density by means of the Langmuir-Schaeffer or Langmuir-Blodgett techniques.[25] However, while these techniques are suitable for depositing monolayers of amphililic materials formed beforehand on the surface of a liquid, they are not appropriate for the deposition of homopolymers or other non-surface-active polymers.

Polymer brushes can be attached to surfaces via physisorption or chemisorption. Physisorption occurs through molecular interactions such as electrostatic,[26] or hydrophobic interactions, whereas chemisorption involves a covalent attachment resulting from a chemical bond being formed between the surface and the polymer. Covalent attachment is usually favored, as it overcomes the disadvantages of thermal- and solvent-driven instability of physisorbed macromolecules. Using covalently attached polymer brushes, changes in polymer conformation upon changes in environmental conditions can be investigated without uncertainties in the grafting density or the surface chemistry. Covalent attachment of polymer brushes can be achieved using both "grafting-to" and "grafting-from" techniques.

Polyelectrolyte brushes can be obtained using similar grafting-to or grafting-from approaches. For example, by using a neutral (hydrophobic)-b-charged (hydrophilic) diblock copolymer, such as polystyrene-b-poly(acrylic acid) and the Langmuir-Blodgett technique, homogeneous polyelectrolyte brushes can be grafted to a surface at a constant and controlled grafting density (Fig. 6.4).[25, 27] Grafting-from has also proven to be a very effective way to create highly dense and well-controlled polyelectrolyte brushes on different surfaces using controlled radical polymerization.[28–30] The extra repulsion provided by the electrostatic and osmotic pressure of the free counterions associated with the charged or ionized chains, compared to neutral polymers, has a significant influence on the normal and frictional forces between two opposing polyelectrolyte layers, as discussed in the following sections.

6.1.3. *Response of neutral polymer brushes to applied load*

The scaling theory of Alexander and de Gennes states that the normal interaction forces between two polymer-brush-covered surfaces (under moderate

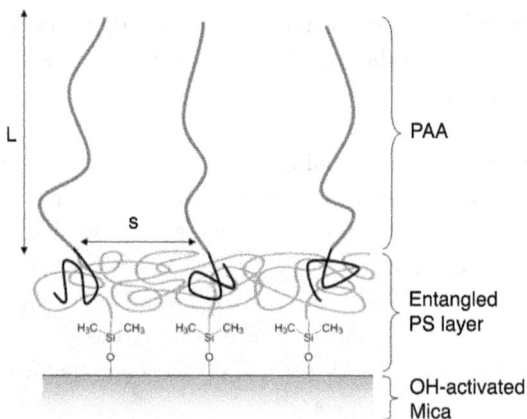

Fig. 6.4. Schematic representation of the PS-b-PAA brush in good solvent with the PS block entangled in a PS monolayer covalently end-grafted on a OH-activated mica surface. s: PAA interchain distance, L: PAA unperturbed brush height. Reproduced from Ref. 25 with kind permission.

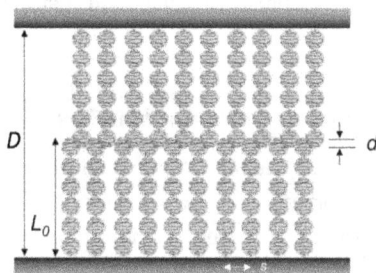

Fig. 6.5. Schematic of tethered polymer brushes, according to the Alexander-de Gennes model (L_0 is the equilibrium height of the brush, s the spacing between adjacent surface-attached polymer chains). Adapted from Ref. 32 with kind permission.

compression) are given as follows[31]:

$$F(D) \cong \frac{k_B T}{s^3} \left[\left(\frac{2L_0}{D} \right)^{9/4} - \left(\frac{D}{2L_0} \right)^{3/4} \right] \qquad (6)$$

in which $F(D)$ is the normal force, T is the temperature, k_B is Boltzmann's constant, s is the spacing between adjacent polymer chains, L_0 is the static polymer length under no compression, and D is the distance between the two substrate surfaces (Fig. 6.5).

The first term in square brackets in Eq. (6) originates from an osmotic repulsion effect as the brushes are compressed, while the second term is due to a free-energy reduction, brought about by compression of the extended chains. The Alexander-de Gennes scaling theory makes the assumption of uniformly stretched polymer chains and an averaged monomer concentration. Milner et al. later extended this into mean-field theory, introducing a parabolic distribution of chain ends, which is more realistic.[33]

Equation (6) only holds as long as a polymer layer is in the "brush" conformation. If in the "mushroom regime", more adhesive interactions are typically observed during compression-decompression cycles. This can be manifested as bridging or hysteresis, and the consequence is generally a lessening of lubricity during shear.[34]

Equation (6) suggests that it is important to characterize the conformation of the polymer layer, both in its compressed and uncompressed state, in order to measure the normal forces between two brush layers in a quantitative manner. The surface forces apparatus (SFA) readily provides such information, since it involves perfectly smooth surfaces and allows absolute measurements of separation, either in a stationary state or during sliding.

6.1.4. Interpenetration of polymer brushes

The fluidity of the interpenetrating polymer chains under compression and shear also plays an important role in determining the very low frictional behavior of polymer-bearing surfaces. Although this was not treated as significant in scaling theory, mean-field theory with the parabolic distribution of chain ends allowed Klein and coworkers to quantify the magnitude of the interpenetration zone (see Fig. 6.5).[35]

$$d \approx s\beta^{1/3} = s(2L_0/D)^{1/3} \tag{7}$$

in which β^{-1} $(= D/2L_0)$ is the compression ratio. The size of the interpenetration zone, d, between two contacting brushes, is not a strong function of compression. The expression can be extended to enable calculation of the shear stress experienced by the chains, while they are being dragged through the interpenetration zone, d.

$$\sigma_s = (6\pi \cdot \eta_{eff} \cdot v_s \cdot \beta^{7/4})/s \tag{8}$$

in which η_{eff} is the effective viscosity of the interpenetration zone, and v_s is the sliding velocity.

The effective friction coefficient can be expressed as

μ_{eff} = (shear force required to slide the compressed brush/normal

load required to compress the brush to separation D)

$$= \sigma_s/F(D) = (6\pi \cdot \eta_{eff} \cdot v_s \cdot s^2 \cdot \beta^{-1/2})/k_B T \qquad (9)$$

in which $F(D)$ is the normal force, as shown in Eq. 6. By comparing with experimental data, Klein has shown the validity of Eq. 8 for the shear forces between polymer-brush layers under moderate compression.[35]

6.1.5. *Polyelectrolyte brushes: Response to applied load and mutual interpenetration*

The effect of the presence of charges within polymer brushes (ionized polymer segments and free counterions) on the conformation and behavior of the brush strongly depends on the nature of the ions, the grafting density, the counterion distribution, and the ionic strength of the medium surrounding the polymer brushes.

Two different polyelectrolyte types (quenched and annealed) and two different added salt regimes (osmotic and salted) are usually considered for the theoretical approaches describing polyelectrolyte brush behavior. The polyelectrolyte can be quenched with a fixed degree of dissociation α or annealed, where α depends on the ionic strength of the medium. The effect of added salt on the polyelectrolyte brush behavior depends on the concentration of added salt C_S compared to the concentration of charged segments ($\alpha\rho$ where ρ is the monomer concentration). The regime where $C_S \gg \alpha\rho$ is called a salted brush. In this regime, the dissociated counterions from the brush can be exchanged with the added ions and the degree of dissociation of the brush is equivalent to that of a free polymer chain in solution. When $C_S \ll \alpha\rho$, the brush is in the osmotic regime, for which ion exchange is suppressed and the degree of dissociation is expected to vary with the grafting density and ionic strength according to the following equation.[13]

$$\alpha \approx \left(\frac{\alpha_s}{1-\alpha_s} \sigma^{-1} (C_{H^+} + C_s) \right)^{1/2} \qquad (10)$$

where α_b is the degree of dissociation of the polymer chain in the solution under the same condition of ionic strength ($\rho_{H^+} + \rho_S$) and σ the grafting density.

Most of the theories developed for expressing the interaction forces between charged brushes are based on the Alexander and de Gennes model and treat the electrostatic contributions assuming a homogeneous monomer distribution throughout the brush and a local electroneutrality inside the brush. They neglect the excluded volume and the rigidity of highly charged chains.[9] These assumptions and simplifications reduce the model complexity and introduce a prefactor into the resulting equations. However, the model developed by Biesheuvel minimizes the simplifications and treats the elastic (chain flexibility), excluded volume, and electrostatic contributions separately for quenched and annealed polyelectrolytes in both osmotic and salted regimes.[14] The resulting interaction energy predicted by the Biesheuvel model is expressed in terms of normalized force $F(D)/R$, which is measured in surface forces experiments between two crossed cylinders of radius R or between a plate and a sphere of radius R. This normalized force is equivalent to the interaction energy between two flat surfaces, according to Derjaguin approximation.[36] Therefore, according to Biesheuvel, as a function of separation distance between two opposing annealed polymer brushes in the osmotic regime (small concentration of added salt), the interaction force is:

$$\frac{F(D)}{R} = 4\pi kT\sigma \left[\left(\frac{3D^2}{2lNa^2} \right) + \left(\frac{N^2\sigma v}{2lD} \right) \right.$$

$$\left. + \left(\frac{N}{m} \ln(1-\alpha) - \frac{2C_S D}{\sigma} \left(\sqrt{1 + \left(\frac{N\sigma\alpha}{2C_S mD} \right)^2} - 1 \right) \right) \right] \quad (11)$$

where a is the segment length, $l = $ Kuhn length$/a$, ν is the excluded volume parameter, and $1/m$ is the fraction of ionizable monomers. The first term on the right-hand side of Eq. 11 corresponds to the conformational contribution (also called elastic), the second one corresponds to the monomer concentration (monomer excluded volume), and the last one expresses the electrostatic contributions (including the electrostatic contribution of polyelectrolyte charges and the osmotic pressure of counter ions). The degree of dissociation for annealed brushes in the osmotic regime is a function of the grafting density, the separation distance, the concentration of added monovalent salt C_s and the pH (or concentration of the proton C_H).[13]

$$\alpha \approx \left(\frac{\alpha_s}{1-\alpha_s} \frac{D}{\sigma} (C_{H^+} + C_s) \right)^{1/2} \quad (12)$$

where α_s is the degree of ionization of the chains in the bulk solution for given C_H and C_s.

In the salted regime, i.e. under the influence of an electrostatic screening effect, the third term (electrostatic contribution) can be simplified, and the equation becomes[13]:

$$\frac{F(D)}{R} = 4\pi kT\sigma \left[\left(\frac{3D^2}{2lNa^2} \right) + \left(\frac{N^2\sigma v}{2lD} \right) \right.$$
$$\left. + \left(\frac{C_S(\alpha\sigma N/2mC_S)^2}{\sigma} \left(\frac{1}{D} \right) \right) \right] \tag{13}$$

Simple scaling laws as well these two last equations have been successfully used to represent the force profiles measured between highly quenched and weakly annealed polyelectrolyte brushes in the osmotic and salted-brush regimes.[24, 25, 37]

Modeling the mutual interpenetration between two opposing charged polymer brushes has been investigated theoretically and by using simulations. The models are more complex than for interpenetration between neutral brushes due to additional parameters, such as the degree of dissociation of the chain and the ionic strength of the medium. Zhulina and Borisov used scaling laws to show that two opposing polyelectrolyte brushes under compression first contract while maintaining a polymer-free gap as the compression increases.[38] Molecular-dynamics simulations have also shown that two opposing polyelectrolyte brushes avoid mutual interpenetration upon compression by folding in upon themselves.[39–41] However, the extent of mutual interpenetration has been shown to depend on the degree of ionization of the chains and the ionic strength of the medium.[40] Significant mutual interpenetration has been predicted in the presence of added salt and the variation of the interpenetration thickness with separation distance, grafting density, and polymer size follows the same scaling law as that observed for two opposing grafted neutral brushes in good solvent. However, the compression between two opposing charged brushes was shown to result in less interpenetration relative to neutral brushes when considering equivalent grafting density and molecular weight. In the presence of mutual interpenetration, the predicted number of segments in the interpenetrated zone is less between charged brushes than it is between neutral brushes under equivalent applied normal forces, because polyelectrolyte brushes remain further apart on account of the electrostatic contributions. Another simulation study has shown that the added salt in the osmotic regime has little effect on the short-range interaction and consequently, the maximum normal force between two planar charged brushes at the point of contact is remarkably unaffected by salt.[42]

6.2. Fundamental Aspects of Lubricating with Polymer Brushes

6.2.1. *Load-carrying capacity under higher compression*

Pressure has a significant effect on the lubricity of surface-grafted polymers, and locally results from a combination of surface roughness, substrate compliance, and the external applied load. It is of great practical interest to determine the limit of pressure at which the highly effective lubricating behavior of polymer brushes is still observable. According to the SFA experimental data in the literature, the regime of *immeasurably* low frictional behavior of polymer brushes seems to be sustainable up to around 1 MPa, depending on the polymer architecture.[43] AFM data on dextran brushes show that this is highly dependent on grafting density.[44] Such low contact pressures are below those present in mammalian articular joints and resemble those between the eyelid and the surface of the eye.

At higher pressures, where friction starts to become measurable, polymer-brush films appear to behave in a glass-like manner as the segment density in the interpenetration zone increases. At $\beta^{-1} = 0.1$, the concentration of polymer is around an order of magnitude higher (45%) than that at zero compression (4.5%).[45] As compression increases further, "stiction" is observed prior to sliding, and, depending on the polymer system involved, detachment of the films can occur, if the compression values are sufficiently high.[37, 43, 53]

6.2.2. *Dependence on shear rate*

As shear rate changes in a polymer-brush-coated contact, it can have a profound effect on polymer conformation and interfacial forces. The relationship between the polymer segment mobility, its relaxation time and the shear rate are the deciding factor.[35, 46, 47]

Although there appears to be some agreement that polymer chains become stretched in the direction of flow while under shear (i.e. diagonally tilted), the effect of shear on polymer brush height and interfacial forces is still a contentious topic.[35, 47–50] The Klein group carried out fundamental

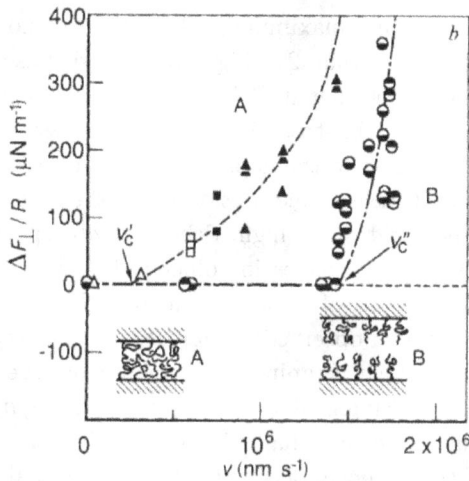

Fig. 6.6. The variation of the increment of $\Delta F_\perp/R$ in the normal force between mica surfaces a fixed distance D apart, bearing terminally anchored, zwitterion-terminated polystyrene layers, as a function of their mean lateral velocity v. Curves A and B are for $D = 94.5 \pm 1$ nm and $D = 155.2 \pm 1$ nm, respectively. From Ref. 46 with kind permission.

experimental studies in this area[46]; when the surfaces were either out of contact or only weakly under compression, an extra force in the normal direction was observed upon shearing at high shear rate (up to several hundred Hz), which would suggest that the brush height of a non-compressed layer was increasing (Fig. 6.6). The shear dependency of this behavior implies that the relationship between the chain relaxation rate and the shear rate is significant.

Many modeling studies have shown consistency with this result.[50] The models are based on decreasing size of the blobs (Fig. 6.5) with shear. In other words the polymer segments' excluded, or screening, volume is reduced due to lateral stretching. This increases the total osmotic interaction of the brush and this is reflected in increasing normal force.[35] Molecular dynamics simulations, however, come up with a different picture[48] in which the vertical chain length of the polymers is essentially unchanged under flow. Grest dismisses many models, since they are based on scaling theory, which cannot serve as a basis for describing shear-rate dependence of brush height.[47] Neutron-reflectivity measurements actually support the absence of brush-height change under shear flows up to 1.3×10^5 Hz.[49] A further complication is the different types of shear used in different experiments: oscillatory in the case of SFA and steady flow in the case of neutron reflectivity.

Klein observes[46] that a maximum in shear-induced normal force occurs when D is slightly greater than $2L$ (Fig. 6.5), decaying as overlap begins to occur, and finally disappearing at $\beta^{-1} \approx 0.4$. This can be readily explained by the unfeasibility of chain stretching under compression, due to the high concentration of monomer in the gap. Under such conditions, with significant compression and measurable friction force, shear thinning appears to dominate when shear rates are high. Schorr *et al.* report shear thinning between two polystyrene brushes in toluene at $\beta^{-1} \approx 0.22$ and 200 nm s^{-1}, while at lesser values of either parameter, Newtonian behavior was observed.[51] Raviv *et al.*[52] observed similar behavior between poly(ethylene glycol) brushes. The shear-thinning behavior was ascribed to the reduced interpenetration zone at high values of shear rate, arising from chain tilting.

It has been suggested that charged polymers slide past each other more easily because of the absence, or small extent, of mutual interpenetration and the presence of hydration layers around the charged polymer segments.[53] In the absence of mutual interpenetration, the shear force between two opposing charged brushes is expected to result from shearing hydrated ions and/or hydrated charged polymer segments, and mainly relies on the low effective viscosity of water confined in thin films. These mechanisms could explain the very low friction coefficients measured experimentally. Molecular-dynamics simulations of the relative shear motion of both neutral and polyelectrolyte end-grafted polymer brushes have shown that polyelectrolyte brushes exhibit smaller friction force than neutral polymer brushes for equivalent grafting density, chain length and applied load.[40,54] Under equivalent normal applied load, the resulting monomer concentration in the interpenetrated layer for the charged brushes is lower than that of neutral brushes. Therefore, a smaller number of segment collisions occur per unit time in the case of charged polymers, resulting in a lower friction coefficient. The friction coefficient is shown to be directly correlated with the interpenetration thickness but more specifically to the number of polymer segments within the interpenetration layer.[40]

6.3. Macro-, Micro- and Nano-Tribological Measurement Approaches for Polymer Brushes

Polymer brushes have been investigated as to their tribological properties on several different scales of length and load, corresponding to the ranges of the atomic force microscope (AFM), surface-forces apparatus

(SFA), microtribometer, and pin-on-disk macrotribometer. The differences between results obtained from these approaches are not only due to scale, but also due to issues such as the relative importance of roughness in the different techniques. In an SFA experiment the roughness (of mica surfaces) is essentially that of atomic corrugation, and therefore the apparent contact area is similar to that of the real contact. This facilitates control of contact pressure, simply by controlling load, which becomes important for determining polymer-brush conformation and forces of interaction. Real and apparent contact areas are also the same for AFM, but for another reason: for a sharp AFM tip, the tip is essentially the size of an asperity and thus it can be considered that it is entirely in contact with the flat surface beneath. As the micro and macro (pin-on-disk) scale is approached, the roughness of the contacts means that multiasperity contact becomes the norm (unless one of the partners is an elastomer) and thus the real contact area is only a fraction of the apparent value. This means that the contact pressure becomes, locally, much higher than the nominal value, and very difficult to control.

Another issue is that of sliding speed. In the case of macroscopic pin-on-disk instruments, a common speed range would be around $\sim 10^{-4}$ to $10^0 \, \mathrm{m \, s^{-1}}$, while for the AFM and SFA it lies in a much lower range of $\sim 10^{-7}$ to $10^{-3} \, \mathrm{m \, s^{-1}}$. This is basically a consequence of the driving mechanisms, which are mechanical motors in the case of pin-on-disk and piezoelectric elements in the cases of AFM and SFA. The speed is of significance for lubricated contacts, since it can determine the degree to which hydrodynamics is important, but in the case of polymers the relationship between the speed and the polymer segment mobility or relaxation time can also play a role in determining interfacial forces.[35, 46, 47]

In a practical sense, exploring polymer-brush-based lubrication on different size scales enables a wide range of different properties to be investigated. Macroscopic investigations are clearly of practical importance, since most lubricated contacts are macroscopic. For example, wear of the brush is really only probed on the macroscopic scale, and it is of the utmost importance in practical applications. Macroscopic pin-on-disk experiments are also particularly suited to testing a wide variety of materials (metals, ceramics, elastomers, thermoplastics) and investigating the effect of macroscopic roughness. Modeling studies of macroscopic contacts are extremely complex, however, due to the multiasperity nature of the interaction. For matching theory to experiment, nanoscopic measurements, such as AFM and SFA are much more appropriate. Such nanoscale approaches allow the mechanisms underlying the efficiency of polymer brushes in reducing friction or shear force between two sliding surfaces to be investigated. Several studies using SFA and AFM have been carried out for different classes of solvated polymer-bearing surfaces, in order to elucidate the role of different parameters, such as polymer conformation, solvent quality, and type of substrate, in reducing friction between surfaces.

6.4. Experimental Studies of Neutral and Charged Systems

6.4.1. *Neutral systems*

While initial studies in the Klein group concerned non-aqueous polymer-brush systems, there were also investigations of quaternary amine-end-functionalized poly(ethylene glycol) (PEG) that had been tethered, via ion-exchange interactions, to opposed mica surfaces in a surface forces apparatus.[52] In comparison to their earlier, non-aqueous studies,[35] the shear forces were found to increase at relatively low compression ratios, this being attributed to a much lower volume fraction in the PEG case. The adsorbed PEG chains were of M_w 3400, and the mean distance between grafting points, s, was calculated as 2.0 nm, corresponding to a coverage, σ, of about 0.25 chains nm^{-2}. In a subsequent paper[55] in the same group, much larger PEG chains (M_w 150 000 or 170 000) were adsorbed onto mica via ion-ligand attachment from an electrolyte solution, attaining a comparable mass coverage, but much lower chain density (due to the higher M_w) than in the previous study. Upon increasing load, the slip plane was observed to shift from chain-chain to chain-mica, resulting in a saturation in friction value.

The first studies of lubrication with PEG brushes under macroscopic conditions were conducted in the Spencer group, who utilized the comb-like copolymer, poly(L-lysine)-g-poly(ethylene glycol) (PLL-g-PEG)[56] to immobilize PEG brushes on oxide surfaces. This copolymer adsorbs spontaneously on negatively charged surfaces at neutral pH values, due to the presence of positive charge on the PLL backbone (Fig. 6.7). In this way, σ values of around 0.5 nm^{-2} of PEG-2000 can be readily achieved.[57]

Both sliding and rolling experiments were carried out between hard (steel and glass) surfaces at pressures up to 0.8 GPa, and the polymer was found to reduce friction by 50% in sliding[59] and up to two orders of magnitude in pure rolling,[60] when added to a buffer solution at approximately 250 ppm. The rolling results can be explained by the much lower shear forces

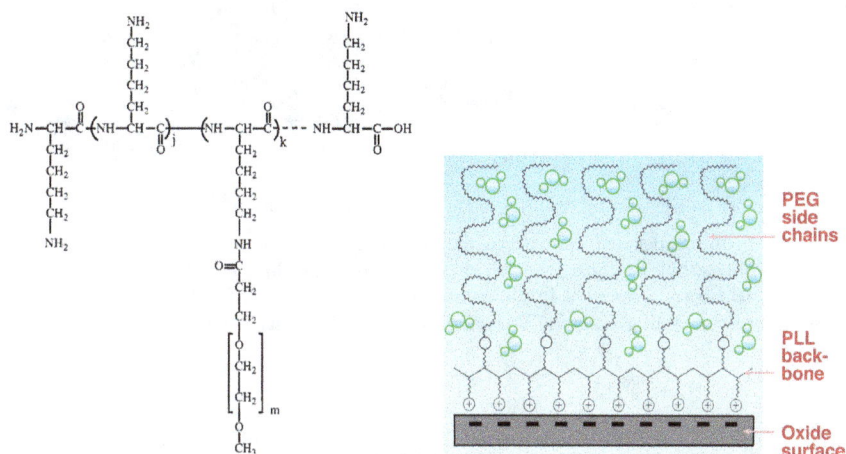

Fig. 6.7. Poly(L-lysine)-*g*-poly(ethylene glycol) molecular structure and a schematic of its interaction with a negatively charged oxide surface. From Ref. 58 with kind permission.

acting in this configuration, and thin-film interferometry measurements in rolling suggested that a brush layer of several nm in thickness was present, even at the lowest speeds investigated.

The electrostatic interaction of the PLL-g-PEG backbone with the oxide is relatively weak, consisting of a sum of small interactions between all unfunctionalized lysines on the backbone and the charges on the oxide. However, it seems to be adequate to maintain the polymer on the surface in the absence of tribological contact, and also weak enough to break under extreme tribological conditions, protecting the integrity of the molecule. Thus, the molecules appear to be removed intact, but only under shear, rapidly being replaced by molecules from the surrounding solution. By labeling surfaces and solution PLL-g-PEG with differently colored fluorescent tags, this "self-healing" property of PLL-g-PEG could be demonstrated (Fig. 6.8).[61]

These macroscopic experiments demonstrate that, during sliding, PLL-g-PEG is continuously removed and replaced, which is a dissipative process. During rolling, where the shear forces are less extreme, this process occurs to a lesser extent. Comparison of the sliding and rolling performance of similar PEG-grafted copolymers with two different binding (backbone) chemistries: PLL and poly(allyl amine) (PAAm) (Fig. 6.9), suggest that the shorter, less flexible NH_3^+-terminated sidechains in PAAm lead to a weaker binding to

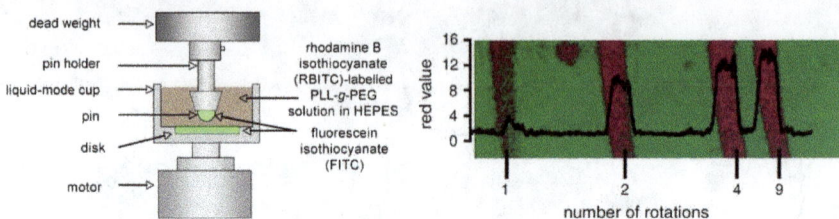

Fig. 6.8. Pin-on-disk experiment to demonstrate the "self-healing" effect of poly(L-lysine)-*g*-poly(ethylene glycol), in which red-fluorescently labeled PLL-*g*-PEG was dissolved in water to serve as a lubricant, after green-labeled PLL-*g*-PEG had been pre-coated onto the sliding surfaces. From Ref. 61 with kind permission.

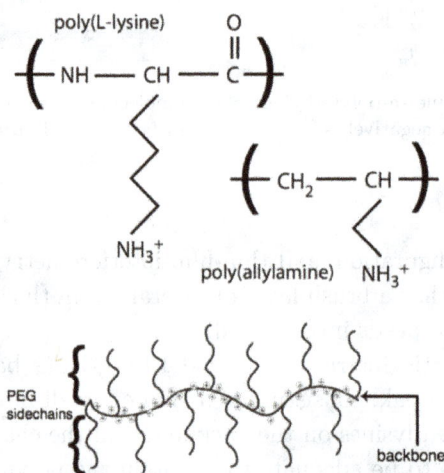

Fig. 6.9. The different backbone structures PLL and PAAm showing the difference and length (and therefore flexibility) in the amine-terminated sidechains. More flexibility translates into greater adsorption and lower susceptibility to shear, resulting in lower friction under sliding conditions. From Ref. 62 with kind permission.

the surface, resulting in both lower coverage and poorer lubrication performance in sliding, while the performance in rolling is essentially unaffected by the choice of backbone.[62]

In order to probe the intrinsic tribological properties of the PEG brush formed from PLL-*g*-PEG, in the absence of the continuous removal and replacement processes, Yan *et al.* investigated the system with atomic force microscopy, using a colloidal, borosilicate probe, both coated and uncoated with polymer, as a countersurface.[63] The results indicated that advantages for lubrication of longer PEG chains and of both sliding surfaces being covered in brushes. Studies of the same system with a surface forces apparatus[64] again highlighted the self-healing behavior: When opposed mica surfaces were coated with PLL-*g*-PEG in buffer solution, no friction could be measured at compressions of up to a 10-fold reduction in the brush thickness. Upon further compression, the friction increased steadily with load until the coating apparently was destroyed. If PLL-*g*-PEG was present in solution, however, the load necessary for film destruction could no longer be reached under the conditions achievable in the SFA.

The comparison of macroscopic and SFA measurements with PLL-*g*-PEG suggest that a common failure mode for the coating (which can however be re-healed, but with a dissipative penalty) involves the high pressures associated with sliding contact between hard asperities. While it is virtually impossible to ensure the absence of asperities (except with the mica surfaces used in SFA), the use of softer materials can reduce the seriousness of the problem. This was illustrated by Lee *et al.*[65] for the case of poly(dimethylsiloxane) countersurfaces, which were rendered negatively charged by means of plasma treatment, and subsequently slid against each other in PLL-*g*-PEG solution (Fig. 6.10).

In the case of the PDMS surfaces slid, untreated, against each other in water, the friction coefficient was very high, due to the hydrophobically induced adhesion between the sliding partners. Plasma treatment rendered the surfaces hydrophilic, but at low speeds, an upturn in the friction coefficient was observed, suggesting that the system was entering a boundary-lubrication regime. In PLL-*g*-PEG solution, this upturn at low speeds was apparently eliminated, due to the replacement of an (elasto) hydrodynamic lubricant film by one whose presence is due to the solvent-enriched layer

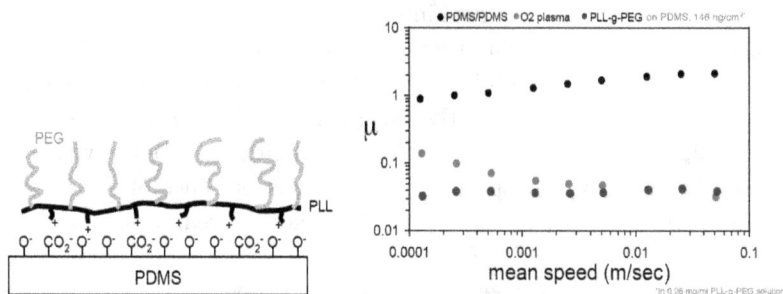

Fig. 6.10. Adsorption of PLL-*g*-PEG onto plasma-treated PDMS, and its effect on friction coefficient. Adapted from Ref. 65 with kind permission.

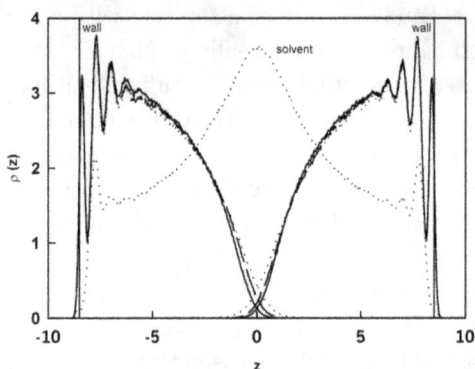

Fig. 6.11. The density profile for both polymer brushes and solvent in a brush-brush sliding system modeled by dissipative particle dynamics. Adapted from Ref. 66 with kind permission.

at the interface of two brushes (Fig. 6.11).[66] This effect has recently been demonstrated even more dramatically for oil-compatible brush systems.[67]

The impact on PEG-brush-based lubrication of the addition of other solvents or non-solvents to water has also been scrutinized. The addition of a poor solvent can lead to an interesting phenomenon in which it apparently remains excluded from the brush at compositions up to high poor-solvent concentrations, at which point the brush rapidly collapses and the friction behavior rapidly transitions between that characteristic of the brush in water to that of the collapsed polymer in the poor solvent (Fig. 6.12). This process, known as *preferential solvation*, was studied by Müller *et al.*[68] by means of optical methods (to obtain the dry polymer mass), a quartz-crystal microbalance (to obtain the mass of polymer including entrained solvent), and colloidal-probe atomic force microscopy (to measure friction).

Under conditions of practical interest, it is often desirable to be able to lubricate with a full fluid film, when speeds and loads permit. With this in mind, a PLL-*g*-PEG-based lubricant, with a higher intrinsic viscosity, was investigated by Nalam *et al.*[69] By using a glycerol-buffer mixture instead of buffer alone, this lubricant was able to show significant improvements in frictional behavior at low speeds, thanks to the presence of the polymer, while at higher speeds, lower friction (and added wear protection) were provided by a fluid film (Fig. 6.13).

PLL-*g*-PEG has also been examined as a potential additive for lubricating polymers and ceramics. In the case of thermoplastics,[1] although the surface charge is generally absent, PLL appears to adsorb by means of

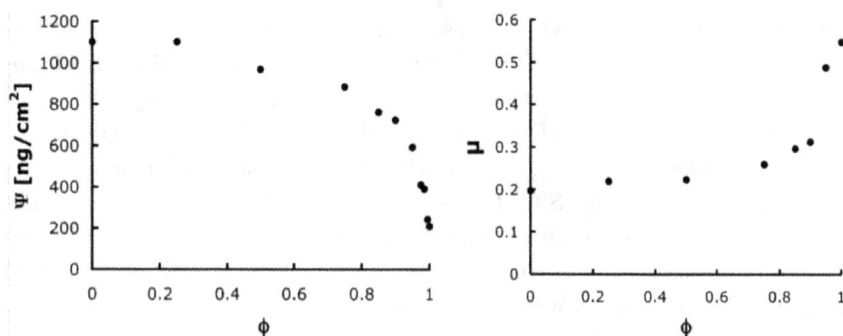

Fig. 6.12. The amount of solvent trapped in the brush, Ψ, per unit area (*left*) and the corresponding friction coefficient measured against a borosilicate bead in AFM (*right*), as a function of the composition of a binary mixture of water and 2-propanol ($\Phi = 1$, pure 2-propanol; $\Phi = 0$, pure aqueous buffer solution). Adapted from Ref. 68 with kind permission.

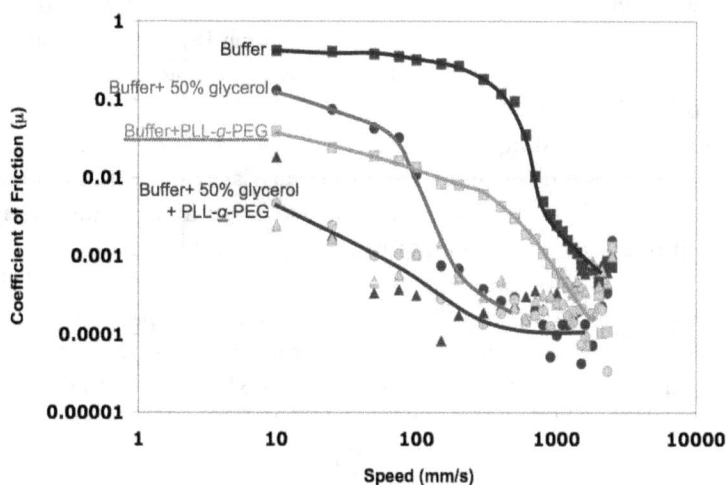

Fig. 6.13. Rolling contact between a steel ball and a glass surface (Load 10N), showing the influence of PLL-*g*-PEG on reducing friction in the boundary region and glycerol on shifting the onset of hydrodynamic lubrication to lower speeds. Adapted from Ref. 69 with kind permission.

hydrophobic interactions between the substrate surface and the four methylene groups in the lysine side-chains, and thus PLL-*g*-PEG leads to substantial (5×) reduction in sliding friction between self-mated polypropylene, even under 10N load. When polymers are slid under PLL-*g*-PEG solution against negatively charged surfaces, such as steel or glass, further advantages are observed. Unlike PEO-PPO block copolymers, which also lower polymer-polymer friction,[71] the additive adsorbs on both sliding partners — both polymer and inorganic surfaces — albeit via different mechanisms.

In the case of ceramics,[72] PLL-g-PEG was investigated as a potential lubricant for Si_3N_4 and SiC. Friction was lowered in the presence of the polymer for SiC under all conditions, but the situation with Si_3N_4 was a little more complicated. In the absence of polymer, Si_3N_4 is known to exhibit high lubricity in water at moderate-to-high speeds, but not at low speeds, provided it is first conditioned by sliding, which has been variously described in the literature as resulting from a smoothening of the surfaces[73] or the formation of a gel-like tribochemical film.[74] The effect of PLL-g-PEG on Si_3N_4 lubrication seems to depend on when it is added. If the ceramic surfaces are first allowed to become conditioned in polymer-free water, then the subsequent addition of PLL-g-PEG appears to extend the low-friction regime to much lower speeds. If, on the other hand, the polymer is already present at the start of sliding, before conditioning is allowed to take place, it appears to prevent the conditioning process that leads to low friction at moderate-to-high speeds.

While PEG-based aqueous lubricant additives have shown interesting tribological properties, a disadvantage of such systems is the oxidational instability of PEG in aqueous environments, even at moderate temperatures.[75] In consequence, attempts were made to replace PEG brushes by dextran brushes, also using a poly(lysine) backbone, and immobilizing it on surfaces in a similar way.[76] Dextran shows similar properties to PEG in this context, PLL-g-dextran lubricating somewhat less well than PLL-g-PEG in sliding geometry (Fig. 6.14). PLL-g-dextran has the advantage of

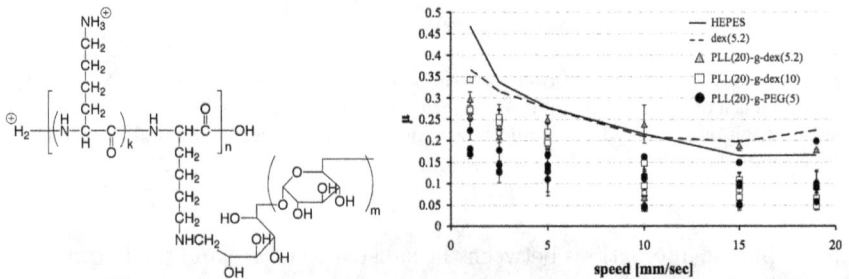

Fig. 6.14. (*Left*) Structure of PLL-g-dextran, (*Right*) Friction coefficient in sliding geometry (steel vs glass, 2N), comparing behavior of a variety of grafting ratios of PLL-g-dextran and PLL-g-PEG. From Ref. 76 with kind permission.

a considerably more straightforward synthesis than PLL-*g*-PEG, since the reducing end of dextran can be used to react directly with the PLL backbone, whereas the PEG synthesis relies on costly end-functionalized PEG. Dextran brushes have the disadvantage, however, of being susceptible to bacterial attack.[77]

The effect of grafting density on friction was investigated for the PLL-*g*-dextran system,[44] a gradient of grafting density being fabricated on a single silicon wafer by means of slow immersion of the wafer into a dilute PLL-*g*-dextran solution. The result was a gradient where the ratio of the spacing between dextran chains to twice the radius of gyration (s/2R$_g$ — see Figs. 6.1 and 6.2) ranged between 2.5 (mushroom regime) to 0.5 (well into the brush regime). Colloidal-probe AFM friction measurements on this surface (Fig. 6.15) revealed that the friction went through a transition from

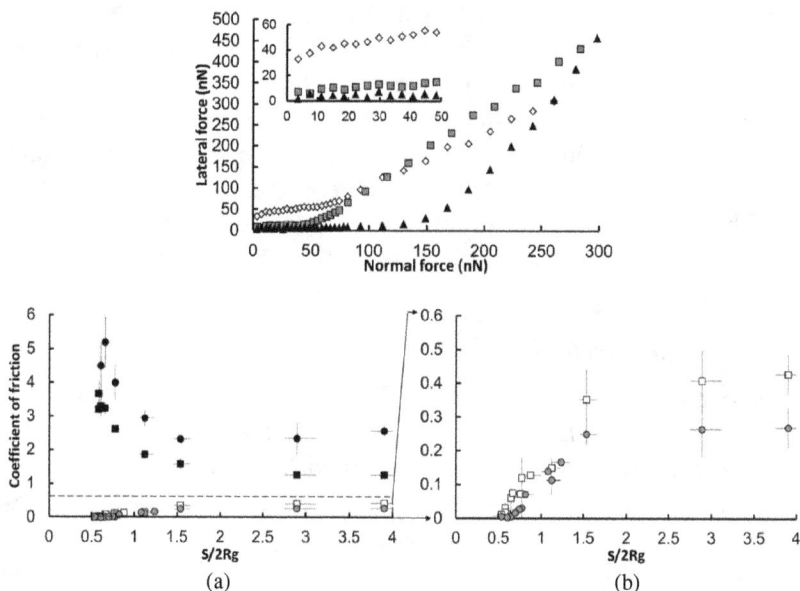

Fig. 6.15. (*Top*) Dependence of lateral force (friction) on normal force at three positions along a dextran chain-density gradient on a silicon wafer for s/2Rg = 2.9 (diamond), 1.2 (square), and 0.6 (triangle). The inset zooms in on the data obtained up to a load of 50 nN. In each case a transition load is visible, at which point the slope changes dramatically. This depends on grafting density. (*Bottom*) (a) Friction coefficient as a function of location along the gradient (i.e. of grafting density, plotted as s/2Rg). Above the dotted line are the data taken at loads higher than the transition load, below (expanded in (b)) are the data taken below the transition load. Squares indicate bare countersurface, circles indicate a countersurface that was itself covered with a dextran brush. From Ref. 44 with kind permission.

low to high as the load was increased. The value of the transition load varied as a function of the chain density. When measuring friction as a function of $s/2R_g$, it was found that below the transition load the friction decreased as $s/2R_g$ decreased towards 0.5 (i.e. the dextran layer became more and more brush-like), whereas above the transition load the opposite trend occurred. This suggests that once the dextran brushes have been significantly compressed, and therefore deprived of coordinating water molecules, they hydrogen bond to the countersurface, significantly increasing the frictional forces measured. This effect becomes greater, as the dextran chain density is increased.

Although grafting-to approaches are highly convenient, grafting-from (methods offer far more flexibility, in terms of grafting density, polymer molecular weight, and block structure. Moreover, grafting-to techniques can readily incorporate crosslinking strategies, bridging the gap between brush and gel lubrication (See Chap. 5). In a study involving photoinifertermediated polymerization (PMP),[78] Li *et al.* used a light-emitting diode (LED) UV source to polymerize acrylamide, adding varying amounts of bisacrylamide, in order to crosslink the brush to different extents (Fig. 6.16).[79] The addition of crosslinker led to both a stiffening and a shrinking of the resulting polymer, as determined by AFM nanoindentation. The friction increased dramatically upon increasing the amount of crosslinker, in the low percentage range, within the polymer layer.

6.4.2. *Charged systems*

The first measurements of the normal and lateral forces between highly charged polymer brushes using the SFA were initiated in Giasson's group. They used a grafting-to approach, with a diblock copolymer, poly(tert-butyl methacrylate)-*b*-poly(glycidyl methacrylate sodium sulfonate) (PtBMA-*b*-PGMAS), on hydrophobized mica surfaces.[24,37,53] The measured range of normal interactions in a salt-free environment, as well as in low ionic strengths, extended well beyond the contour length of the polyelectrolyte chain, suggesting the presence of an effective charged plane at the polymer/water interface.[24] The shear or friction forces measured upon sliding the charged polymer brushes were extremely low, even when the brushes compressed down to a volume fraction close to one.[37,53] Effective friction coefficients were lower than about 0.0006–0.001, even at low sliding velocities and at pressures of up to several atmospheres (typical of those in living systems). On further compression, at separation distances corresponding

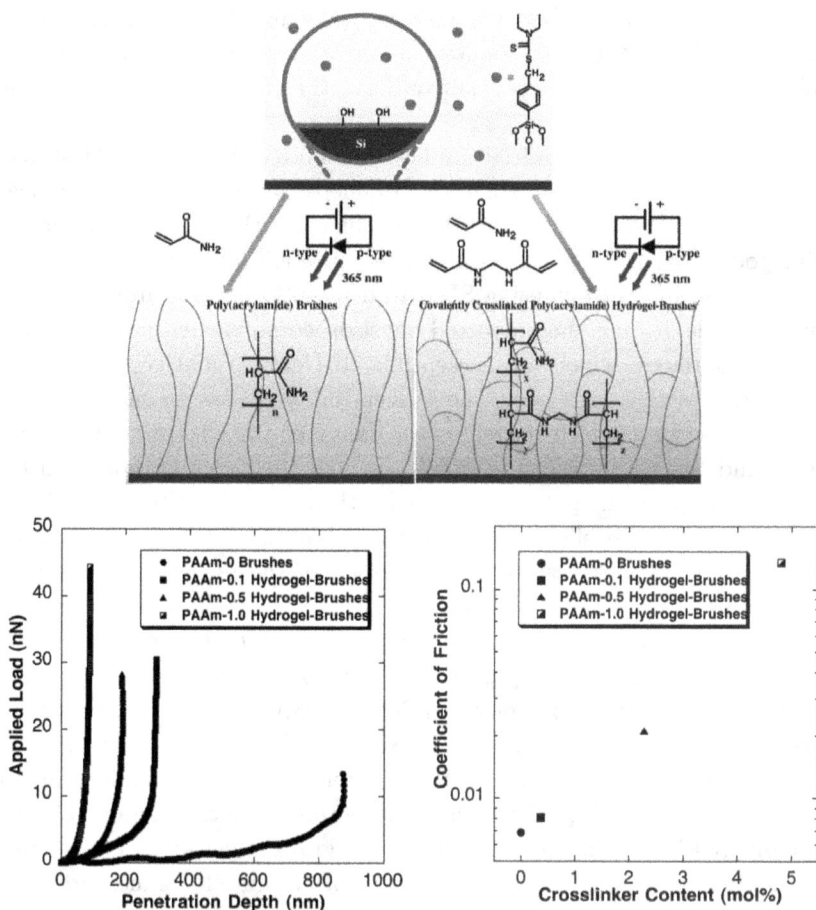

Fig. 6.16. (*Top*) Synthetic approach to poly(acrylamide) brush-gels, involving silane-functionalized photoiniferters, a 365 nm UV LED source, and bisacrylamide to crosslink the growing brushes. (*Lower Left*) The effective modulus of the brush/brush-gels increases and the swollen height decreases, upon the incorporation of increasing amounts of crosslinker. All polymer coatings have comparable dry thicknesses, around 39($\pm$6) nm, "x" in the notation PAAm-x referring to the percentage of crosslinker in the feed. (*Lower Right*) The friction coefficient, measured in a microtribometer against an oxidized PDMS surface, increases as the amount of crosslinker in the polymer layer increases. From Ref. 79 with kind permission.

to pressures of a few atmospheres, the shearing motion led to the abrupt removal of most of the chains from the gap between the surfaces, associated with a sharp and significant increase in friction force. The extremely low frictional forces between the charged brushes (prior to their removal) was attributed to the exceptional resistance to mutual interpenetration displayed by the compressed, counterion-swollen brushes, together with the fluidity of the hydration layers surrounding the charged, rubbing polymer segments.

Following Giasson's initial studies, numerous SFA and AFM studies of surface forces between different polymer-bearing surfaces have been reported in the literature, in order to elucidate the mechanisms underlying good lubricity of polyelectrolytes in aqueous-based systems. However, most studies carried out using SFA until recently suffered from a significant obstacle, in that they reported physisorbed polymers (or macroinitiators) on surfaces, which are susceptible to cleavage and slipping at the polymer/substrate interface. The susceptibility is due to the screening effect of the ions, which decreases the range of the electrostatic interactions, and the mobility of the small ions and their competition with the charged groups of the polyelectrolytes for the adsorption sites at the interface. Both of these effects decrease the strength of the polyelectrolyte-interface interactions, leading to desorption of the polyelectrolyte layers. This is also true for neutral polymer brushes in good solvent that are attached to mica through ionic bonding between the anchoring group and mica, such as the poly(vinylpyrrolidone)-b-polystyrene diblock and end-terminated polystyrene polymer with $N^+(CH_3)_2(CH_2)_3SO_3^-$ as terminating group. Brush detachment was observed for both polymers when the friction force reached about $100 \, \mu N$.[80,81] Therefore, conclusive understanding on the role of polymer conformation and ionization on the friction behavior is hampered by the inability to control the grafting density. The fact that mica, which is the most reliable substrate for accurate force measurements using SFA, generally exposes chemically inert surfaces, has in the past essentially precluded covalent, stable polymer attachment. To overcome this weakness, reliable protocols were developed in 2007 by Liberelle *et al.*, allowing surface-reactive sites (silanol groups) to be created reproducibly on mica, enabling covalent attachment of silane-functionalized molecules and polymers.[23,83]

The first studies of nanolubrication using the SFA with charged brushes without uncertainties regarding the slip plane were conducted in the Giasson group, who used the copolymer, poly(styrene)-b-poly(acrylic acid) (PS-b-PAA), irreversibly attached to functionalized mica surfaces[25] as illustrated in Fig. 6.17. At a controlled and constant polymer grafting density, the effect of the degree of dissociation of the polymer, tuned by the ionic strength of the medium surrounding the brushes, on the overall

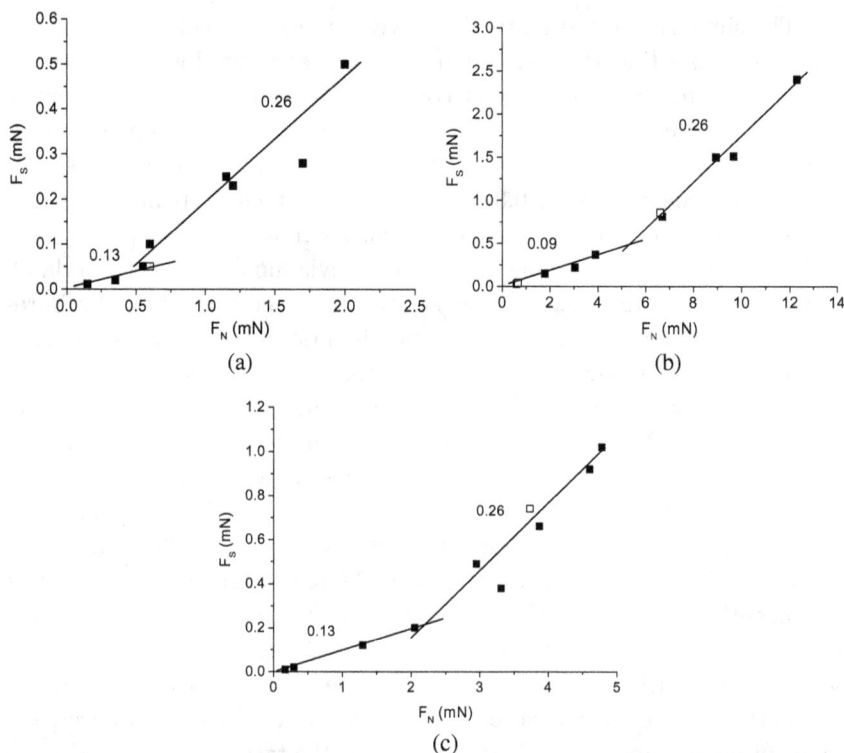

Fig. 6.17. Friction force F_s as a function of the normal force (load) F_N between two opposing PAA brushes across water sliding against each other at a constant sliding velocity $(1 \,\mu\text{m.s}^{-1})$ over a traveled distance of $15 \,\mu\text{m}$. At pH 5.5, (a) without added salt, (b) 10 mM NaCl and (c) 100 mM NaCl. The PAA grafting density σ is $4.5 \,\text{nm}^2.\text{chain}^{-1}$. From Ref. 25 with kind permission.

interaction forces were investigated. The reversibility and reproducibility of the normal-force profiles and friction measurements confirmed the irreversible attachment of the PAA brushes to the mica substrate. The experimental interaction-force profiles agree very well with scaling models developed for neutral and charged polymer brushes.[14,31] The variation in the friction force measured between the two PAA brushes in motion with respect to each other as a function of applied normal load appeared to be non linear over the entire range of normal forces studied $(0 < F/2\pi R < 100 \,\text{mN/m})$ but similar for the different degrees of polymer ionization — including the neutral case. Low friction coefficients (ca 0.1) were measured for the low-load regime $(F/2\pi R < 5 \,\text{mN/m})$, whereas higher values (0.3) were observed for larger applied loads.

The similarity in the F_s vs F_N curves for the different degrees of ionization suggests that the increase in friction is associated with an increase in mutual interpenetration upon compression, as observed with neutral polymers. Indeed, the effect of the PAA charges and added ions was more significant on the repulsive normal forces than on the friction forces. The high friction coefficients (ca 0.3) were measured at relatively high pressures (40 atm) with no surface damage or polymer removal.

A similar study, involving poly(2-(dimethylamino)ethyl methacrylate), PDMAEMA, grafted from gold substrates, has been carried out by Nordgren and Rutland using AFM. This was the first investigation of the nanotribology of a dual-responsive system that enabled the effect of both degree of ionization and solvent quality to be studied, without changing the adsorbed amount.[84] A variation of pH was used to modify the degree of ionization of the brushes. The measured friction force varied linearly with the applied load within a range of applied normal force ($0 < F/2\pi R < 1\,\mathrm{mN/m}$) — smaller than that used for the PAA study using the SFA ($0 < F/2\pi R < 100\,\mathrm{mN/m}$). Moreover, the friction coefficient significantly decreased (from ca 1 to ca 0.1) as the degree of ionization of the brushes increased (from pH 11 to 3 at 24°C). At first glance, these results seem to contradict the results obtained for the PAA brushes, which suggest no effect of degree of ionization on the friction coefficient. However, the range of normal applied loads is different for the two techniques (SFA and AFM) and in the low-load regime ($F/2\pi R < 1\,\mathrm{mN/m}$), the friction coefficient is similar (≈ 0.1) for both charged brushes, PAA and PDMAEMA. However, the friction coefficient measured at pH 11, where the amine groups are expected to be neutral and the polymer in good solvent conditions, is relatively high compared to neutral brushes under similar conditions of load and sliding velocity. Other factors, such as surface roughness, may be involved and could explain this discrepancy.

In a study involving PMP, Heeb et al.[85] used an LED UV source to polymerize methacrylic acid, up to estimated molecular weights of $>10^5$. Friction values, measured in pin-on-disk sliding against an oxidized PDMS countersurface at macroscopic loads, were considerably lower than those measured for PLL-g-PEG-functionalized surfaces, dropping below the detection limit of the tribometer ($\mu < 0.005$). Over 1000 sliding cycles could be performed without any apparent increase in friction, as long as the molecular weight of the polymer was of order 10^5 Da. Lower molecular weights (e.g. 14 kDa) showed degradation after several hundred cycles (Fig. 6.18).

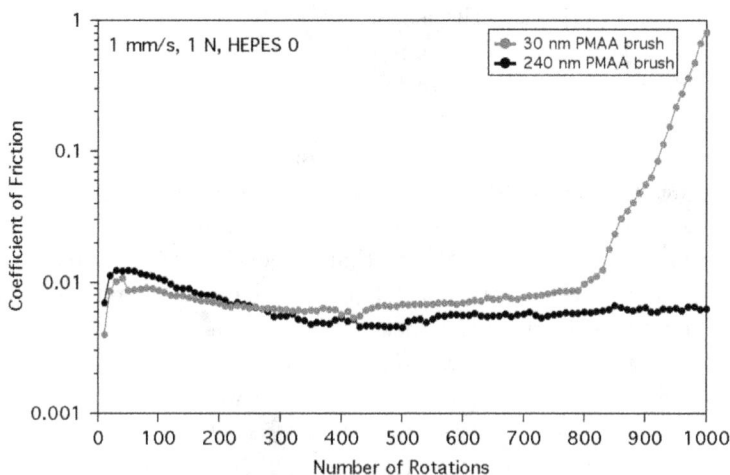

Fig. 6.18. Long-term pin-on-disk experiments involving poly(methacrylic acid) brushes with two distinct brush heights (30 and 240 nm — approximately 14 kDa and 110 kDa, respectively). The shorter brush (gray circles) did not sustain the applied tribological stress (1 N normal load, 1 mm/s sliding speed against oxidized PDMS in HEPES buffer) for 1000 rotations, while the 240 nm PMAA samples displayed very low friction coefficients throughout the entire experiment. From Ref. 85 with kind permission.

6.4.3. *Wear of brushes*

Polymer brushes prevent wear of the substrate beneath by providing a cushion that separates hard asperities from interacting with each other. Providing the cushion is thicker than the asperities, even when the brush is under compression, no substrate wear will occur. However, this picture is complicated by locally high pressures at asperities, and this is where brush wear presumably initiates, upon tribological stress.

The two brush-wear possibilities are chain scission or grafting failure. In the case of grafting-to systems, chains that are covalently or strongly bonded to a surface, such as PEG-silanes on silica[61] or PAA on mica[25] are clearly more resistant to grafting failure in an aqueous environment than chains that are held in place by physical interactions (e.g. electrostatic bonding — in PLL-*g*-PEG or hydrophobic interactions in — PtBMA-*b*-PGMAS on hydrophobized mica). The latter derive their stability towards simple desorption from the cooperative molecular interactions with the surface (by the multiple protonated amine groups or hydrophobic groups), but nevertheless are susceptible to removal upon shearing. In the case of silane bonds with the substrate, despite its stability, the bond will ultimately break under asperity-asperity shear, never to reform. The presence of silanes in the surrounding medium would not improve the situation, since they would be hydrolytically unstable and would polymerize in solution. In the case of PLL-*g*-PEG, the relative ease by which the polymers can be sheared, presumably intact, from the surface, facilitates their replacement by molecules from solution, and leads to the self-healing phenomenon described above. Thus, in the case of PLL-*g*-PEG, the wear of the substrate is determined by the amount of time that it spends in an unprotected state, i.e. ultimately by the kinetics of diffusion to the surface and readsorption of the polymer. Chain scission has not been directly determined during wear, but it seems likely that stronger surface bonding is more likely to lead to chain scission as a failure mode, since weaker bonding will favor intact removal of the polymers.

Another factor has recently come to light in the context of oil-based brush lubrication, and this is the viscosity of the solvent. Solvents within brushes can exhibit extremely high effective viscosities, as has been observed previously by Feiler *et al.*[86] and this can contribute to a further, time-dependent, cushioning behavior of the brush. Comparing the durability of oil-compatible grafting-from brushes in a number of chemically similar lubricating oils of different viscosities, it was clear that the wear of brushes (and underlying substrate) was significantly reduced in the presence of more viscous oils.[67]

Crosslinking may be one approach to improving the robustness of brushes, although, as mentioned above, there is some evidence where crosslinked brushes display higher friction than those with no crosslinking. Preliminary results from the Giasson and Drummond laboratories suggest that adding covalent cross-link bridges to an adsorbed polyelectrolyte layer leads to significant changes in the tribological behavior, which are strongly dependent on the applied load and sliding amplitude. Cross-linking the polymer layer appears to increase its cohesion and wear-resistance under small deformations, but increases its propensity to catastrophic damage under large deformations.

Preliminary results from the Spencer laboratory suggest that covalent crosslinking of brush chains at a location close to the substrate actually worsens durability towards wear. This can be rationalized as follows: if wear at the substrate-polymer interface is considered to be a fracture process, it may be necessary to enhance the viscoelastic dissipation at this interface, in order to inhibit crack propagation at the polymer-substrate interface, and thus lower wear. These arguments have

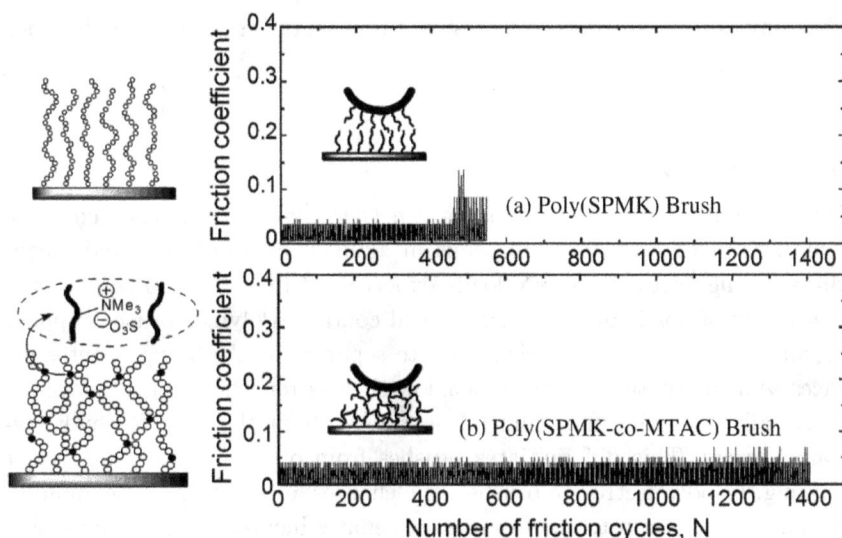

Fig. 6.19. Poly(3-sulfopropyl potassium methacrylate (SPMK) (top), and Poly(3-sulfopropyl potassium methacrylate (SPMK)-co-2-(methacryloyloxy) ethyltrimethylammonium chloride (MTAC)) (poly(SPMK-co-MTAC) (bottom) brushes slid against each other along a reciprocating distance of 10 mm at a speed of 1.5×10^{-3} mm/s at a load of 0.49 N in water at 298 K. The presence of electrostatic crosslinks appears to enhance the brush-wear resistance. Reproduced with kind permission from Ref. 88.

been previously been raised by Ahagon and Gent,[87] who observed that for adhesion of polymers to glass "the strength of adhesion decreased with increasing crosslinking". The group of Takahara,[88] however, have copolymerized positively and negatively charged monomers (Fig. 6.19), which then crosslink themselves electrostatically: The polymer poly(3-sulfopropyl potassium methacrylate (SPMK)-*co*-2-(methacryloyloxy) ethyltrimethylammonium chloride (MTAC)) (poly(SPMK-*co*-MTAC) was grafted from a glass surface, and then slid against a similar countersurface. The wear properties appear interesting, the system apparently displaying low friction over 1400 cycles at a pressure of 139 MPa. When poly(SPMK) only was used, the system was far less resistant towards brush wear. While the observation of low friction and low wear for a crosslinked system may appear contradictory to the poly(acrylamide) results mentioned above, it is possible that weaker, electrostatic crosslinks are beneficial, in comparison to strong covalent crosslinks, such as those in the poly(acrylamide) case, and

this may again suggest that a dissipation mechanism needs to be in place
to impart wear resistance.

6.5. Conclusion

Water-compatible polymer brushes, whether charged or uncharged, have
been shown to be highly lubricious in aqueous environments under conditions ranging from the nN loads experienced in AFM and SFA to the
Newton-scale loads of macrotribological contact. Polymer brushes appear
to function by means of their ability to separate hard, sliding countersurfaces while presenting a solvent-rich, low-shear-strength inter-brush layer —
essentially providing fluid-film lubrication without the need for hydrodynamic forces. This differentiates brushes from other varieties of polymer
coating. In polyelectrolyte brushes, the charges appear to provide an additional entropic term to the free energy penalty incurred upon compression
and interpenetration, thereby enhancing the ability of the brush to separate
the surfaces.

The main challenge facing polymer brushes appears to be resistance to
wear. To some extent this issue can be sidestepped, however, if polymer systems are used that can spontaneously re-establish a lubricating layer following removal. This does, however, lead to a penalty in friction. Crosslinking
may be a possible solution to the wear issue for non-regenerable brushes,
although it is not yet clear whether it can really bring benefits. Of the
approaches tried to date, reversible crosslinking appears to be a promising
avenue to pursue.

References

1. S. Lee and N. D. Spencer, *Science* 319, 575–576 (2008).
2. P. J. Flory, *Principles of Polymer Chemistry*, Cornell University Press
   (1953).
3. N. D. Spencer, *Tailoring Surfaces*, World Scientific (2011).
4. T. Wu, K. Efimenko, P. Vlcek, V. Subr and J. Genzer, *Macromolecules* 36,
   2448–2453 (2003).
5. A. Halperin, M. Tirrell and T. P. Lodge, *Adv Polym Sci* 100, 31–71 (1992);
   Polymers at Interfaces, G. J. Fleer, M. A. Cohen Stuart, J. M. H. M. Scheut-
   jens, T. Cosgrove, B. Vincent, Chapman & Hall, London (1993).
6. O. V. Borisov, F. A. M. Leermakers, G. J. Fleer and E. B. J. Zhulina, *Chem
   Phys* 114, 7700 (2001).

7. R. Israels, F. A. M. Leermakers, G. J. Fleer and E. B. Zhulina, *Macromolecules* 27, 3249 (1994).
8. E. B. Zhulina, J. K. Wolterink and O. V. Borisov, *Macromolecules* 33, 4945 (2000).
9. P. Pincus, *Macromolecules* 24, 2912 (1991).
10. M. W. Matsen, *Eur Phys J E* 34, 45 (2011).
11. R. Israels, F. A. M. Leermakers and G. J. Fleer, *Macromolecules* 27, 3087–3093 (1994).
12. R. Israels, F. A. M. Leermakers, G. J. Fleer and E. B. Zhulina, *Macromolecules* 27, 3249–3261 (1994).
13. E. B. Zhulina, T. M. Birshtein and O. V. Borisov, *Macromolecules* 28, 1491 (1995).
14. P. M. Biesheuvel, *J Coll Interf Sci* 275, 97 (2004).
15. R. C. Advincula, W. J. Brittain, K. C. Caster and J. Rühe, *Polymer Brushes: Synthesis, Characterization, Applications*, Wiley-VCH (2004).
16. M. Ejaz, S. Yamamoto, K. Ohno, Y. Tsujii and T. Fukuda, *Macromolecules* 31, 5934–5936 (1998).
17. J. Chiefari, Y. K. Chong, F. Ercole, J. Krstina, J. Jeffery, T. P. Le, R. T. A. Mayadunne, G. F. Meijs, C. L. Moad, G. Moad, E. Rizzardo and S. H. Thang, *Macromolecules* 31(16), 5559–5562 (1998).
18. Z. Bao, L. B. Merlin and G. L. Baker, *Macromolecules* 39, 5251–5258 (2006).
19. C. Devaux, J.-P. Chapel, E. Beyou and P. Chaumont, *Eur Phys J E* 7, 345–352 (2002).
20. M. Ejaz, S. Yamamoto, K. Ohno, Y. Tsujii and T. Fukuda, *Macromolecules* 31, 5934–5936 (1998).
21. B. Lego, M. Francois, W. G. Skene and S. Giasson, *Langmuir* 25, 5313–5321 (2009).
22. M. Himmelhaus, T. Bastuck, S. Tokumitsu, M. Grunze, L. Livadaru and H. J. Kreuzer, *Europhys Lett* 64, 378–384 (2003).
23. B. Liberelle and S. Giasson, *Langmuir* 23, 9263–9270 (2007).
24. T. Abraham, S. Giasson, J. F. Gohy and R. Jérôme, *Langmuir* 16, 4286–4292 (2000).
25. B. Liberelle and S. Giasson, *Langmuir* 24(4), 1550–1559 (2008).
26. G. L. Kenausis, J. Vörös, D. L. Elbert, N. P. Huang, R. Hofer, L. Ruiz, M. Textor, J. A. Hubbell and N. D. Spencer, *J Phys Chem B* 104, 3298–3309 (2000).
27. E. P. K. Currie, A. B. Sieval, M. Avena, H. Zuilhof, E. J. R. Sudholter and M. A. C. Stuart, *Langmuir* 15, 7116 (1999).
28. R. Barbey, L. Lavanant, D. Paripovic, N. Schuwer, C. Sugnaux, S. Tugulu and H. A. Klok, *Chem Rev* 109, 5437–5527 (2009).
29. B. Lego, W. G. Skene and S. Giasson, *Macromolecules* 43, 4384–4393 (2010).
30. W. He, H. Jiang, L. Zhang, Z. Cheng and X. Zhu, *Polym Chem* (2013).
31. S. Alexander, *J Phys (France)* 38, 983–987 (1977); P. G. De Gennes *Macromolecules* 13, 1069–1075 (1980).

32. S. Lee and N. D. Spencer, Ch. 21 in *Superlubricity*, Ed. A. Erdemir and J.-M. Martin, Elsevier (2007).

33. S. T. Milner, T. A. Witten and M. E. Cates, *Macromolecules* 21, 2610–2619 (1988); S. T. Milner, Z. G. Wang and T. A. Witten, *Macromolecules* 22, 489–490 (1989).

34. J. Klein and P. Luckham, *Nature* 308, 836–837 (1984); U. Raviv, R. Tadmor and J. Klein, *J Phys Chem B* 105, 8125–8134 (2001).

35. J. Klein, *Ann Rev Mater Sci* 26, 581–612 (1996).

36. B. Derjaguin, *Kolloid-Z.* 69, 155 (1934).

37. U. Raviv, S. Giasson, N. Kampf, J. F. Gohy, R. Jérôme and J. Klein, *Langmuir* 24, 8678–8687 (2008).

38. E. B. Zhulina and O. V. Borisov, *J Chem Phys* 107, 5952 (1997).

39. O. J. Hehmeyer and J. M. Stevens, *J Chem Phys* 122, 134909 (2005).

40. M. Sirchabesan and S. Giasson, *Langmuir* 23, 9713 (2007).

41. C. Ibergay, P. Malfreyt and D. J. Tildesley, *Soft Matter* 7, 4900 (2011).

42. M. W. Matsen, *Eur Phys J E* 35, 13 (2012).

43. J. Klein, E. Kumacheva, D. Perahia and L. J. Fetters, *Acta Polymer* 49, 617–625 (1998).

44. K. J. Rosenberg, T. Goren, R. Crockett and N. D. Spencer, *ACS Appl Mat Interf* 3, 3020–3025 (2011).

45. J. Klein, E. Kumacheva, D. Malahu, D. Perahia and L. J. Fetters, *Nature* 370, 634–636 (1994).

46. J. Klein, D. Perahia and S. Warburg, *Nature* 352, 143–145 (1991).

47. G. S. Grest, *Adv Poly Sci* 138, 149–183 (1999).

48. P. Y. Lai and K. Binder, *J Chem Phys* 98, 2366–2375 (1993); P. Y. Lai and C. Y. Lai, *Phys Rev E* 54, 6958–6961 (1996); L. Miao, H. Guo and M. J. Zuckermann, *Macromolecules* 29, 2289–2297 (1996).

49. D. Nguyen, C. J. Clarke, A. Eisenberg, M. H. Rafailovich, J. Sokolov and G. S. Smith, *J Appl Cryst* 30, 680–683 (1997); S. M. Baker, G. S. Smith, T. C. Anastassopoulos, A. A. Vradis and D. G. Bucknall, *Macromolecules* 33, 1120–1122 (2000).

50. J. L. Barrat, *Macromolecules* 25, 832–834 (1992); V. Kumaran, *Macromolecules* 26, 2464–2469 (1993); J. L. Harden and M. Cates, *Phys Rev E* 53, 3782–3787 (1996); M. Aubouy, J. L. Harden and M. E. Cates, *J Phys II France* 6, 969–984 (1996).

51. P. A. Schorr, T. C. B. Kwan, S. M. Kilbey II, E. S. G. Shaqfeh and M. Tirrell, *Macromolecules* 36, 389–398 (2003).

52. U. Raviv, J. Frey, R. Sak, P. Laurat, R. Tadmor and J. Klein, *Langmuir* 18, 7482–7495 (2002).

53. U. Raviv, S. Giasson, N. Kampf, J. F. Gohy, R. Jérôme and J. Klein, *Nature* 425, 163–165 (2003).

54. Y. Ou, J. B. Sokoloff and M. J. Stevens, *Phys Rev E* 85, 011801 (2012).

55. L. Chai and J. Klein, *Macromolecules* 41, 1831–1838 (2008).

56. G. L. Kenausis, J. Vörös, D. L. Elbert, N. P. Huang, R. Hofer, L. Ruiz, M. Textor, J. A. Hubbell and N. D. Spencer, *J Phys Chem* 104, 3298–3309 (2000).

57. S. Pasche, S. M. De Paul, J. Vörös, N. D. Spencer and M. Textor, *Langmuir* 19, 9216–9225 (2003).
58. N. P. Huang, R. Michel, J. Vörös, M. Textor, R. Hofer, A. Rossi, D. L. Elbert, J. A. Hubbell and N. D. Spencer, *Langmuir* 17, 489–498 (2001).
59. S. Lee, M. Müller, M. Ratoi-Salagean, J. Vörös, S. Pasche, S. M. De Paul, H. A. Spikes, M. Textor and N. D. Spencer, *Tribol Lett* 15, 231–239 (2003).
60. M. Müller, S. Lee, H. A. Spikes and N. D. Spencer, *Tribol Lett* 15, 395–405 (2003).
61. S. Lee, M. Müller, R. Heeb, S. Zürcher, S. Tosatti, M. Heinrich, F. Amstad, S. Pechmann and N. D. Spencer, *Tribol Lett* 24, 217–223 (2006).
62. W. Hartung, T. Drobek, S. Lee, S. Zürcher and N. D. Spencer, *Tribol Lett* 31, 119–128 (2008).
63. X. Yan, S. S. Perry, N. D. Spencer, S. Pasche, S. M. De Paul, M. Textor and M. S. Lim, *Langmuir* 20, 423–428 (2004).
64. T. Drobek and N. D. Spencer, *Langmuir* 24, 1484–1488 (2008).
65. S. Lee and N. D. Spencer, *Tribol Inter* 38, 922–930 (2005).
66. D. Irfachsyad, D. Tildesley and P. Malfreyt, *Phys Chem Chem Phys* 4, 3008–3015 (2002).
67. R. Bielecki, M. Crobu and N. D. Spencer, *Tribol Lett* 49, 263–272 (2013).
68. M. T. Mueller, X. Yan, S. Lee, S. S. Perry and N. D. Spencer, *Macromolecules* 38, 3861–3866 (2005).
69. P. C. Nalam, J. N. Clasohm, A. Mashaghi and N. D. Spencer, *Tribol Lett* 37, 541–552 (2010).
70. S. Lee and N. D. Spencer, *Lubrication Science* 20, 21–34 (2008).
71. S. Lee, R. Iten, M. Müller and N. D. Spencer, *Macromolecules* 37, 8349–8356 (2004).
72. W. Hartung, A. Rossi, S. Lee and N. D. Spencer, *Tribol Lett* 34, 201–210 (2009).
73. J. G. Xu and K. Kato, *Wear* 245, 61–75 (2000).
74. H. Tomizawa and T. E. Fischer, *Tribol Trans* 30, 41–46 (1987).
75. T. V. Kryuk, V. M. Mikhal'chuk, L. V. Petrenko, O. A. Nelepova and A. N. Nikolaevskii, *Pharma Chem J* 36, 32–35 (2002).
76. C. Perrino, S. Lee and N. D. Spencer, *Tribol Lett* 33, 83–96 (2009).
77. N. Argibay, C. Perrino, M. Rimann, S. Lee and N. D. Spencer, *Lubrication Science* 21, 415–425 (2009).
78. T. Otsu, M. Yoshida and T. Tazaki, *Makromol Chem Rapid Commun* 3, 133–140 (1982).
79. A. Li, E. M. Benetti, D. Tranchida, J. N. Clasohm, H. Schönherr and N. D. Spencer, *Macromolecules* 44, 5344–5351 (2011).
80. P. A. Schorr, T. C. B. Kwan, S. M. Kilbey, E. S. G. Shaqfeh and M. Tirrell, *Macromolecules* 36, 389–398 (2003).
81. J. Klein, E. Kumacheva, D. Perahia and L. J. Fetters, *Acta Polymerica* 49, 617–625 (1998).
82. B. Liberelle, X. Banquy and S. Giasson, *Langmuir* 24, 3280–3288 (2008).
83. B. Lego, W. G. Skene and S. Giasson, *Langmuir* 24(2), 379–382 (2008).
84. N. Nordgren and M. W. Rutland, *Nano Letters* 9, 2984–2990 (2009).

85. R. Heeb, R. M. Bielecki, S. Lee and N. D. Spencer, *Macromolecules* 42, 9124–9132 (2009).
86. A. Feiler, M. A. Plunkett and M. W. Rutland, *Langmuir* 19, 4173–4179 (2003).
87. A. Ahagon and A. N. Gent, *J Polym Sci Polym Phys Ed* 13, 1285–1300 (1975).
88. M. Kobayashi, M. Terada and A. Takahara, *Faraday Discussions* 156, 403–412 (2012).

Chapter 7

Water-Like Lubrication of Hard Contacts by Polyhydric Alcohols

Jean Michel Martin* and Maria Isabel De Barros-Bouchet
Laboratory of Tribology and System Dynamics
University of Lyon-Ecole Centrale de Lyon, 69134 Ecully, France
*jean-michel.martin@ec-lyon.fr

7.1. Introduction

First, let us consider friction on ice as a model to examine the exact role of water in achieving low friction. Water is known to be involved in the slipperiness of ice and particularly in skiing and skating. The friction coefficient measured during actual speed skating is as low as 0.005 but, at very low sliding speed, there is an appreciable increase in the coefficient of friction. Theories about the presence of water between the rubbing surfaces are focused on the formation of water by (i) pressure melting, (ii) melting due to frictional heating, and (iii) on the liquid-like properties of the ice surface. Originally, the classic explanation was attributed to the pressure that could lower the freezing point of water significantly. The contact pressure calculated in skating has been often claimed to cause ice to melt, thus reducing the friction between skate and ice. A simple calculation using skates 3-mm-wide and 20-cm-long gives a total pressure of about 105 Pa (0.1 MPa or 10 atm) leading to a reduction of the melting temperature of ice by less than 1°C. Later, slipperiness of ice was also attributed to the quasi-fluid water layer that coats any ice surface, providing a permanent lubricant. However, at −20°C, the ice drastically decreases in slipperiness so it just resembles any other solid. Friction increases markedly as the temperature falls and at −80°C, it is five or six times greater than it is at 0°C.[1] Under −160°C, the layer is as little as one molecule thick. Other works support the view that frictional heating can produce local surface melting and therefore a low value of friction.[2] Some of us have studied friction on low-temperature ice at high contact pressure.[3] Using a UHV tribometer, we deposited a thin layer of ice (about 10-nm-thick) on a steel disk and then friction tests were performed on this layer in the temperature range of $-140°C$ to $-100°C$ (133–173 K). The average contact pressure calculated was about 130 MPa. We observed that friction decreases with the increasing temperature linearly from 0.1 to 0.05, demonstrating that ice might be lubricated at relatively low temperatures. A simple theoretical analysis indicated that a frictionally melted layer was responsible for the low friction experimentally observed. Obviously, in the case of ice and whatever the mechanism of water release, the water is present *in situ* in the contact zone and it is provided by melting of the ice material itself. From this brief discussion on the slipperiness of ice, it is clear that a thin layer of water is able to lubricate the contact (even at relatively high pressure) only if it is generated within the interface under the combined effect of pressure and shear and also, presumably, by frictional heating.

Water has a low viscosity, so it cannot directly form an EHL film in a sphere-on-plane contact at high contact pressure, even at high sliding speed. Thus, water is not a good lubricant for hard steel contact. In the hydrodynamic lubrication regime when a complete fluid water layer is formed, a friction coefficient of 0.002 has been reported in the literature.[4] However, liquid water (as well as water vapor) is known to have some ability to lubricate diamond and amorphous carbon (ta-C and a-C) materials in the boundary regime at high contact pressure and low sliding speeds. The reason for this is not definitively understood and is still under debate but it is generally attributed either to a graphitization of carbon and subsequent water-molecules intercalation, or to the occurrence of some tribochemical reactions between water and nascent carbon atoms. Recently, Carpick[5] has shown by XANES analyses that the passivation of the diamond surface by OH termination (and not graphitization) was certainly at the origin of its low friction in the presence of water vapor. In another recent work, Morita *et al.*[6] have used MD calculations to show that sliding between two OH-terminated diamond surfaces yields a very low friction coefficient (below 0.01).

It is also interesting to notice that some ceramic materials such as SiC, Si_3N_4, Al_2O_3 are also well known to be lubricated by water and that tribochemical reactions lead to smooth surfaces and subsequent water hydrodynamic lubrication.[7,8] However in these latter cases, high wear rate due to corrosion leads to a drastic decrease of the contact pressure and the

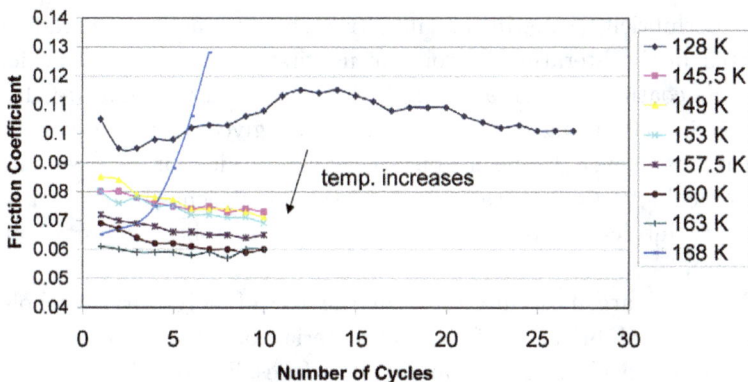

Fig. 7.1. Friction coefficient of steel on ice in ultrahigh vacuum at different temperatures. It is observed that the friction decreases as the temperature increases. This is attributed to the formation of a quasi-liquid water layer by ice melting in the contact interface.

occurrence of the HD lubrication is probable. We suspect that hydroxylation of these surfaces can also explain the good tribological performances in water. Unfortunately, there is no data with alcohol lubrication of these ceramics and more work is necessary. Here, we are interested in the boundary lubrication by polyols at high contact pressures and low sliding speeds in hard-contact situations.

7.2. Polyhydric Alcohols and Carbohydrates as Lubricants

Selective transfer was the name given in the 1970s to various phenomena involving lubrication by polyhydric alcohols and particularly glycerol $C_3H_8O_3$ and copper-based materials and ferrous surfaces as friction pairs. It was widely reported to produce extremely low friction (superlubricity) in the technical literature of the former USSR. Basically, the mechanism was attributed to the formation of a beneficial soft metallic film, possibly copper, which has an ability to reduce friction. Later, in the 1990s, the role of colloidal and gel-like products, which result from the tribochemical reaction, was emphasized and the rheology of the colloidal film was thought to play the key role in the friction reduction. At this point, the interest of Western tribologists in these ideas was negligible, mainly due to the ambiguous definition of the selective transfer mechanism and also difficult access to Russian literature. However, a recent article on this matter was published in 2006 by Ilie.[9]

Superlubricity experiments in ultrahigh vacuum have already been reported in the literature by some of us, first in 1993 using stoichiometric MoS_2 coatings,[10] in 1996 using hydrogenated carbon films a-C:H, and recently in 2007 using ta-C carbon films and glycerol.[11] However, superlubricity has never yet been approached for lubrication of steel surfaces in practical situations. Steady-state friction coefficients below 0.04 under boundary lubrication of steel samples have never been reported in the literature.

Here, we are most interested in the use of polyhydric alcohols (or *polyols*) for the lubrication of different material combinations used in practical situations involving carbon and steel surfaces. Figure 7.2 shows chemical formulae of well-known polyols and particularly cyclic hexols, "cyclitols" such as phenol, resorcinol, pentaerythritol, and myo-inositol. In the past, these polyols have been widely used to synthesize the corresponding esters as lubricant base oils. However, no attempt has been made to evaluate the intrinsic tribological properties of such polyhydric alcohols. In this work, we will focus on glycerol and *myo*inositol which is classified as a member of the vitamin B complex. It is often referred to as vitamin Bh in vitamin and health-supplement guides, although it can be synthesized by the human body. The chemical formula of *myo*inositol is $C_6H_{12}O_6$. In its most stable geometry, the inositol ring is in the chair conformation. There are nine stereoisomers, all of which may be referred to as inositol. However, the natural isomer has a structure in which the first, third, fourth, fifth, and sixth hydroxyls are equatorial, while the second hydroxyl group is axial.

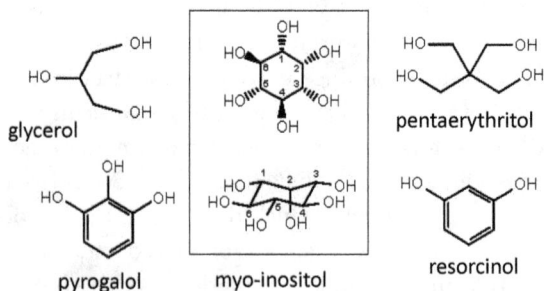

Fig. 7.2. Chemical formulae of some polyhydric alcohols: glycerol, pyrogallol, resorcinol, pentaerythritol, and *myo*-inositol — also known as vitamin Bh. All these molecules are small and have different shapes such as spherical, crab shape or spider shape. They do not contain aliphatic chains, distinguishing them from most of the amphiphilic molecules used as traditional lubricant additives.

All these polyol molecules, including glycerol, are relatively small and have different shapes such as spherical, crab shape or spider shape, etc. Unlike most of the amphiphilic molecules used in traditional lubricants, they do not contain aliphatic chains. In this chapter, we show and discuss tribological results obtained with glycerol on steel and diamond-like coatings (DLCs).

7.3. Lubrication of Steel by Glycerol

7.3.1. *Glycerol under severe/mixed regime*

In the testing conditions with glycerol using steel surfaces with a composite roughness of about 15 nm, a severe mixed/boundary lubrication regime is thought to be achieved. The value of the film thickness using EHL theory in a sphere-on-plane configuration was calculated according to Harmrock and Dowson formulation.[12] More accurately, we calculated the λ ratios (theoretical EHL film thickness divided by the composite surface roughness) at a typical contact pressure of 0.8 GPa, at three temperatures 25, 40, and 80°C and different sliding speeds up to 5.0 mm.s^{-1}. Glycerol has a viscosity of 0.934 Pa.s and a piezoviscosity coefficient of 5.4 GPa^{-1} at 25°C, 0.283 Pa.s and 5 GPa^{-1} at 40°C, and 0.028 Pa.s and 4.3 GPa^{-1} at 80°C, respectively.

Figure 7.3 represents the evolution of the λ parameter as a function of sliding speed for the three temperatures under consideration. For example, using a sliding speed of 1 mm.s^{-1} at ambient temperature (25 C), λ is below 0.4 (see intersection between dashed line in Fig. 7.3), indicating a severe mixed/boundary lubrication regime with a possible EHL glycerol layer average thickness of about 9 nm. It is to be noticed that there is no abrupt separation between ultrathin EHL, mixed and boundary regimes when the surfaces are highly polished.

A friction test performed with pure glycerol at ambient temperature with a sliding speed of 1 mm/s and a contact pressure of 800 MPa is presented in Fig. 7.4(a). Average evolution of friction coefficient and electrical contact resistance (ECR, in logarithmic scale) are presented. To obtain average evolutions of friction and ECR (logarithmic scale) as a function of the number of cycles (Fig. 7.4(a)), we neglected the part of data coming from the extremities of the friction track when the sliding speed is not constant and changes in direction, and we calculated the average value for each passage (one cycle corresponds to two passages). Any details of these evolutions at each passage and position (e.g., fluctuations) can be easily

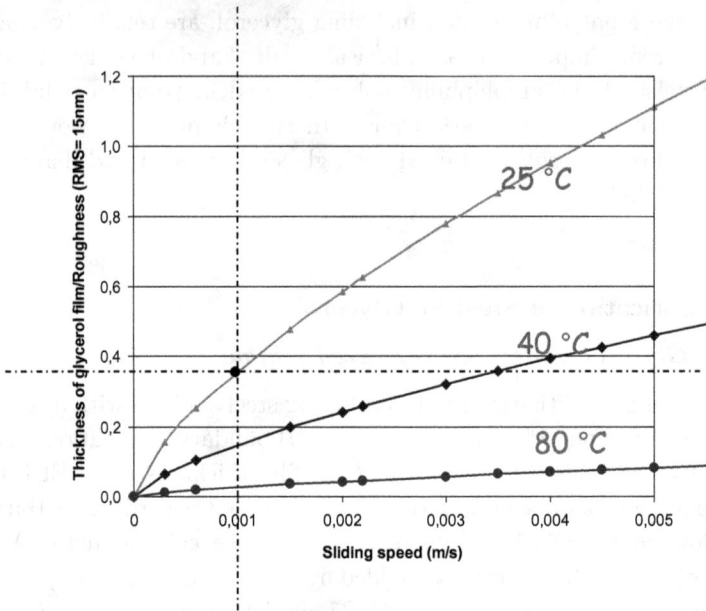

Fig. 7.3.   Calculation of film thickness in the EHL regime. Evolution of the λ parameter for three temperatures, 25, 40, and 80°C, as a function of the sliding speed for a contact pressure of 800 MPa and a sliding speed of 2.5 mm/s. With kind permission of Ref. 15.

obtained from the retrospective analysis of the raw data when it is needed. At the beginning of the test, friction is very low (0.01) and the contact resistance is very high (typically 10^6 ohms). During 400 cycles (about 33 min test duration), friction remains low but the contact resistance tends to decrease significantly and progressively, reaching 10^4 ohms after 400 cycles. Then friction starts to increase slightly as ECR decreases drastically. After 800 passes, when the ECR becomes negligible, then friction suddenly increases to values around 0.1, which is a typical value for boundary lubrication of steel surfaces in presence of polar molecules. For comparison, similar tests previously performed with a PAO base oil gave much higher friction (0.2–0.25) and very low ECR. For each test, we measured the following values: F_{min} corresponding to the lowest value of friction at the beginning, D_{ECR} values during this period, and Nc the number of passes with the lowest friction (say below 0.04). ECR is useful to study the presence of an insulating tribofilm inside the contact area. Thus, we correlate the value of friction coefficient to the value of ECR. Fig. 7.4(b) plots the friction coefficient (linear scale) as a function of ECR values (log scale) during the

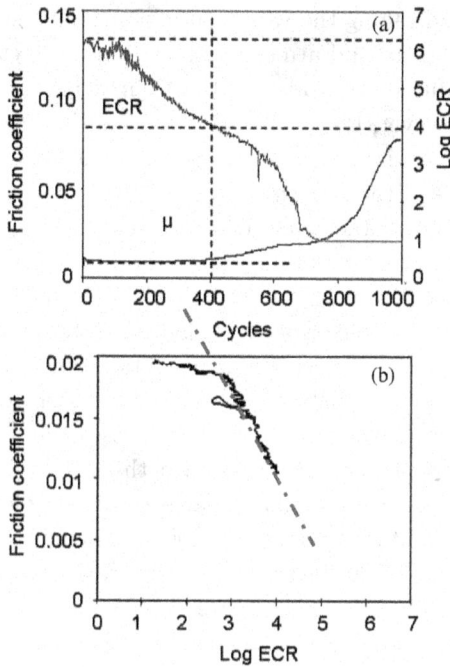

Fig. 7.4. Friction test performed with pure glycerol at ambient temperature with a sliding speed of $1.0 \, \text{mm.s}^{-1}$ and a contact pressure of $800 \, \text{MPa}$. a) Average evolution of friction coefficient and electrical contact resistance (ECR, in logarithmic scale) are presented. b) Evolution of the friction coefficient versus the electrical resistance. With kind permission of Ref. 15.

low friction period. A straight line is obtained when the insulating film is present, which suggests an exponential decay of the contact resistance as a function of thickness of a tribofilm inside the contact area. After stopping the test, we observed the wear scars on the sphere by optical microscopy. During the low friction period (below 200 cycles), the wear track is nearly not visible on both the sphere and the flat. However, when friction drastically increases after 500 cycles, the wear scar is clearly visible on the sphere, indicating a collapse of the glycerol tribofilm and metal contacts. The origin of the contact resistance observed in Fig. 7.4 can be attributed to the existence of a homogeneous glycerol layer in the Hertzian contact area, the diameter of which is about 50 micrometers. Pure glycerol, like pure water, is not an electrically conducting fluid. When it is anhydrous, its electrical conductivity is about $5.59 \, 10^{-6} \, \text{S/m}$ at a temperature of $20°\text{C}$. If we consider the ohmic resistance of a cylinder of glycerol ($50 \, \mu\text{m}$ diameter and 6 nm thick) and taking the value of the electrical resistivity of glycerol at ambient temperature and atmospheric pressure, the calculated value of the electrical conductivity is about $10^{-7}\,\Omega$, in good agreement with the measurements. However, the explanation for the simultaneous decrease of ECR and increase of friction is not straightforward. This can be attributed either to a decrease of the film thickness and/or to a chemical degradation of glycerol inside the contact area. If we plot the friction coefficient (linear scale) as a function of ECR values (log scale) during the superlow friction period (see Fig. 7.4(b)), we obtain a straight line in the low friction regime on two ECR decades (10^2–10^4 ohms). This suggests the existence of some tunnelling conductivity in the Hertzian contact when the film thickness becomes below 5 nm. This behavior was already observed in the literature by Mansot,[13] for example, using a similar friction test with steel surfaces and an ester lubricant. The existence of a thin EHL glycerol layer completely separating the two metallic surfaces in the very low friction regime is confirmed by the fact that no wear scar is observed by optical microscopy. But after the increase of the friction coefficient, wear is observed, suggesting that the protective glycerol layer progressively collapses. This thin layer of glycerol itself is found able to give an anomalous low value of friction. This is in agreement with a previous work by Vergne,[14] who found similar traction values in full EHL lubrication of steel surfaces with glycerol. Actually, the viscosity of glycerol is high enough to allow an EHL film to be established, but its very low pressure-viscosity coefficient prevents high traction values, in contrast to paraffin, for example, which has a solid-like behavior in an EHL contact. At 80°C, the measured friction coefficient remains equal to 0.12 all through the test. At this temperature, the λ parameter is below 0.1 (see Fig. 7.3) and the glycerol film thickness is certainly not able to completely separate the steel surfaces to avoid metallic contacts. These results suggest that a low friction coefficient is obtained when a glycerol film is present inside the contact area. We have no clear explanation as to why friction increases after a certain time (or cumulative sliding distance), which is not negligible (typically a few hundred of cycles or some tens of minutes duration). The most probable explanation is that the glycerol molecules are progressively dissociated inside the contact under the effect of pressure and shear and liberate water and some degradation products such as acids and aldehydes. We previously observed some corrosion patterns in the Hertzian contact area when the test was stopped after a certain time.[15] We attributed this fact to the degradation of glycerol into water molecules and acid or aldehyde fragments and this was further confirmed by MD calculations in the case of DLC-coated surfaces.[16] It is known that mixtures of glycerol and water become rapidly less viscous, even at a few percent water concentration. The decrease in bulk viscosity of glycerol due to water generation could lower the film thickness too much, and metallic contacts would begin to occur. Moreover, corrosion of steel by acids can change the surface roughness and this can give an additional effect on the λ ratio. However, the effect of *in situ* water formation in the contact could also be an advantage, decreasing the shear value of glycerol at high contact pressure. In this regard, the effect of pressure could provide more shear-induced water molecules in the contact area only and consequently better friction reduction. It is expected that this low quantity of water produced inside the contact area during a test would not immediately affect the water concentration and the viscosity in the bulk lubricant volume providing the EHL film thickness. To study this point in more detail, additional tribological tests were performed with glycerol containing 10 wt% of water (not shown here). A friction coefficient of 0.05–0.06 was obtained during 100 cycles, afterwards it rapidly increased to 0.12, indicating that the presence of water in glycerol drastically reduces the performance of the lubricant.

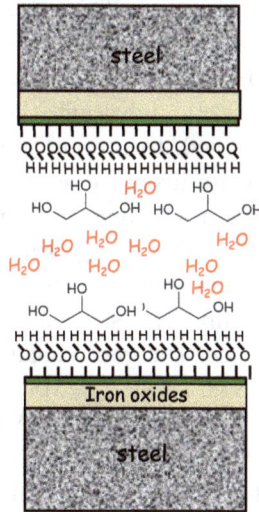

Fig. 7.5.   Schematic representation of the chemistry of the interface under boundary lubrication of a steel surface by glycerol molecules. The water molecules originate from the friction-induced dissociation of glycerol. Friction is governed by a hydrogen bond network. With kind permission of Ref. 15.

To summarize, glycerol is certainly dissociated in the contact area at high pressure, generating confined water molecules which could be a benefit for achieving low friction, but water is progressively released into the glycerol at the outlet of the contact, lowering its bulk viscosity and consequently the EHL film thickness. Probably, acid and/or aldehyde fragments can be neutralized by glycerol when they leave the contact zone. An alternative, but not exclusive, mechanism is the increase of steel surface roughness due to the generation of aggressive species and the subsequent corrosion of the surface, exposing carbide grains as asperities.

7.3.2. *Glycerol under EHL regime*

Traction measurements have been performed with a ball-on-disk tribometer.[17] The two specimens are driven by independent motors to produce the desired slide-to-roll ratio with high precision and stability. They are made from bearing steel and they have been carefully polished to present very smooth surfaces. The bottom of the ball dips into a reservoir containing 20 ml of glycerol, ensuring fully flooded conditions in the contact. The contact, the lubricant and the two shafts that support the specimens are thermally isolated from the outside and are kept at constant temperature by an external thermal control system. This assembly was designed to limit heat transfer from or towards the contact zone, leading to experimental conditions as close as possible to those considered in the numerical model. The experimental volume (i.e. the lubricant, the ball specimen and lower face of the disk) is also confined, to prevent any air flow from the room environment. All parts that are likely to be in contact with the lubricant were previously cleaned using a 3-solvent procedure to ensure the absence of any chemical contamination. During experiments, we collected four different samples of used glycerol during the test at different sampling times for experiment 1 and two samples for experiment 2. The volume collected is about 1 ml in each instance. Different chemical analyses of the liquid were performed including NMR, IR, GC/MS and also Karl Fischer for water detection. NMR analyses (BRUKER AVANCE 400) were performed at 30°C with a QNP probe ^{1}H/^{13}C/^{19}F/^{31}P of 5 mm. The ^{1}H NMR spectra (dilution in deuterated acetone) for pure glycerol (sample no. (1)) shows two peaks, one at 2.31 ppm from the 5 protons of the CH and CH2 groups and a second one at 3.87 ppm corresponding to protons of the hydroxyl group (OH) in the glycerol molecule, (the ^{13}C NMR spectra were very similar for all samples). The peak at 2.31 ppm does not show any change in the carbon skeleton of glycerol during the friction test. This result is confirmed by ^{13}C NMR spectra, which show no change after the test (not shown here) and we observe the increase of a new peak in the ^{1}H NMR spectrum at 3.36 ppm. This new line at 3.36 ppm is attributed to labile protons, because we observe a shift when the spectrum is acquired at a higher temperature. This could be tentatively be attributed to water molecules. Other ^{1}H NMR spectra (not shown) were also recorded after dilution of glycerol in deuterated trichloromethane (CDCl3): under these conditions, we observed a new peak at 8 ppm, possibly attributed to an aldehyde chemical group. Karl Fischer analyses show a slight increase in water content in glycerol from 0.36% to 1.24% for experiment 1 and from 0.62% to 3.25% for experiment 2. Thus we can conclude that we observe a tribochemical dissociation of the glycerol as a function of the duration of the friction test, although it is very limited in quantity (a few per cent of molecules are eventually dissociated by the friction process in the whole lubricant volume).

Chemical analyses of the glycerol lubricant after friction experiments show its partial tribochemical decomposition into water and other species. This result is reinforced by the difference observed between the friction coefficients measured and calculated at high velocity (1 m/s), the calculated friction being twice higher than measured value. This fact provides strong evidence that glycerol is dissociated during its transit through the friction interface and that the water produced inside the contact zone reduces the traction force. Knowing the dimensions of the contact area (assuming a Hertzian diameter of 330 μ m), the sliding speed and the film thickness, we can easily evaluate the quantity of lubricant going through the contact in a given time. For example, in the case of 1 m/s sliding speed, a calculated film thickness of 300 nm, we found 0.36 ml per hour. For 0.01 m/s and a film thickness of 14 nm, the quantity of lubricant becomes only 1.7 10^{-4} ml/hour. During our test (1 hour at 1 m/s and 20 hours at 0.01 m/s), the total quantity of glycerol that passed through the contact is about 0.4 ml compared to the 20 ml contained in the reservoir and dissociation comes mainly from the first step in the high-speed regime (1 m/s). This indicates that the increase of the water content of 1% for the entire glycerol quantity (which approximately corresponds to 0.2 ml) does not reflect the water content that is effectively produced in the contact zone. A rough estimation of the glycerol/water conversion produced in the 300-nm-thick EHL film is at least 40%. At this stage it is interesting to consider the viscosity of aqueous solutions of glycerol that are reported in the literature.[18] At operating temperature (50°C), the viscosity of glycerol is about 142 mPas but it

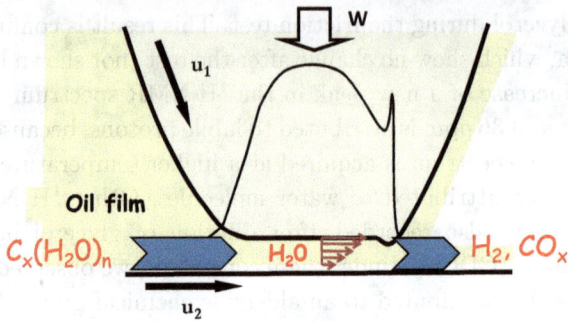

Glycerol viscosity is 1400 mPl at 25 °C
Gly (80%)/water (20%); Viscosity is 60 mPl
Gly (50%)/water (50%); Viscosity is 6 mPl

Fig. 7.6. Schematic representation of friction-induced dissociation of carbohydrates under the combined effects of pressure, shear and temperature. The dissociation can involve only a part of the molecule (glycerol for example) and smaller fragments of acids or aldehydes are also formed.

drastically decreases to 3.8 mPas when 40% of water is added. However, the homogeneity of the water/glycerol mixture is certainly not constant over the entire experimental volume. Due to the design of the tribometer and the relatively high viscosity of the initial glycerol, the water produced within the contact is likely to remain close to the sphere location. Thus the fluid that arrives at the contact inlet has a reduced viscosity that first induces a thinner film. Clearly, both the decrease of the viscosity at the inlet and inside the EHL film thickness can contribute to the lower friction coefficient that we observed in the test compared with the calculation conducted with pure glycerol.[17]

7.4. Lubrication of Diamondlike Carbon by Glycerol

Figure 7.7 shows a summary of the steady-state friction data obtained for different material combinations (steel, a-C/H and ta-C coatings) with different tribometers, and in three different selected environments at 80°C: (i) ultrahigh vacuum @ 10 nPa partial pressure, (ii) in the presence of liquid pure glycerol, and (iii) in the presence of glycerol+1 wt % of myo-inositol (*cis*-1,2,3,5- *trans*-4,6-cyclohexanehexol). The superlubricity of *a*-C:H coatings under ultrahigh vacuum is observed and has previously been reported by some research groups, including some of the present authors.[19]

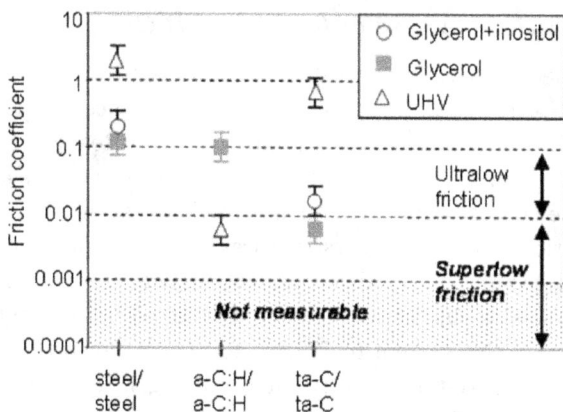

Fig. 7.7. Steady-state friction coefficients values obtained with steel/steel, a-C:H/a-C:H, and ta-C/ta-C friction pair materials, lubricated in three different environments. All tests were performed at a temperature of 80°C. The friction tests in liquid pure glycerol and glycerol+1 wt% *myo*-inositol were performed at 270 MPa contact pressure and 0.3 m s^{-1} sliding speed. Superlubricity is achieved for the first time with ta-C in the boundary lubrication regime with glycerol, approaching pure rolling. With kind permission of Ref. 16.

Although ta-C has been little studied in detail in the literature, we obtained astonishing friction results — approaching superlubricity — for the ta-C/ta-C combination under boundary lubrication conditions with pure glycerol at 80°C. As shown in Fig. 7.7, the friction coefficient of the lubricated ta-C friction pairs is drastically lower than the friction of both a-C:H/a-C:H and steel/steel combinations. These results strongly suggest that the superlow friction behavior involves the interaction between the ta-C coating material and the alcohol groups contained in the glycerol molecule. As a result, a very thin and low-shear-strength tribofilm is formed on the ta-C sliding surface.[20] It appears that superlubricity is directly related to the alcohol chemical functions (−OH) of the glycerol molecule. However, as shown in Fig. 7.7, in the case of ta-C at 80°C, the addition of myo-inositol does not significantly improve the friction level.

7.5. Evidence of Water Formation by Computer Simulation

We present here molecular-dynamics simulations for a model tribological system using ta-C as surface materials, glycerol, and hydrogen peroxide as lubricants, particularly focusing on the role of water generated in the friction interface.

7.5a. Atomistic structure of bulk ta-C DLC

We used molecular dynamics (MD) simulations to study the friction of bare and OH-terminated ta-C surfaces in the presence of one to four layers of glycerol molecules.[16] MD studies used the ReaxFF reactive force field based on quantum-mechanical calculations of structures and chemical reactions. Previous studies have documented the accuracy of ReaxFF to describe reactions and to prepare structures for amorphous systems of many carbon-based systems. To obtain an amorphous carbon structure from simulations, we started with a periodic cell containing 512 atoms in the diamond structure, heated this system to 7722°C for 2 ps to form the liquid phase, then quenched the system to 27°C at a cooling rates of 1127°C/ps and −173°C/ps, and finally equilibrated the structure at 27°C for 3 ps. The process was carried out for densities ranging from 2.7 to 3.4 g/cm^3. The most stable DLC corresponds to 3.0–3.3 g/cm^3. We found that lower cooling rates decrease the energy of the carbon structure leading to more stable ta-C structures with densities ranging from 3.0 to 3.3 g/cm^3. For the density of experimental value 3.24 g/cm^3, we found that 83% of the carbon atoms have sp^3 character, whereas 16.6% have sp^2 character and 0.4% atoms have sp^1 character. We found that all of the sp^3 atoms are interconnected to form a percolating tetrahedral network, to which isolated sp^2 atoms or sometimes short, branched chains of sp^3 atoms attach. The sp^1 atoms sometimes lie between two sp^2 atoms (allene) and sometimes are isolated. The simulation model of bulk ta-C is consistent with XANES and TEM-EELS experimental results. Therefore, the ta-C model with 3.24 g/cm^3 density and 83% of sp^3 carbon (close to the ta-C material in the experimental part, see Fig. 7.7) was used in the following sliding simulation.

7.5b. Atomistic structure of the surface for bare and OH-saturated ta-C

To construct the surface of ta-C, we cut through the cell with planes perpendicular to each of the x, y, and z directions. For each direction, we attempted to cut the solid at 30 places, equally spaced at 0.5 Å and spanning the periodic supercell of 14.75 Å. In each case, the two parts of the ta-C were first separated by 1.5 Å and the positions of the atoms were optimized to minimize the energy. Then we selected the best plane from each of the three sets and carried out an MD simulation at 27°C while separating the planes by a total of 2.5 Å. Then we selected the best of these cases as the actual surface of this particular ta-C structure. For the ta-C surface slab, there are

52% sp^3, 37% sp^2, and 11% sp^1 atoms. About 0.4% of the surface atoms have one bond connected with bulk C atoms. The main result from this calculation is an enrichment in sp^2 and sp^1 hybridized carbon at the top surface of the ta-C material compared with the bulk, which is consistent with the experimental TEM-EELS observations on practical surfaces. To saturate the surface with OH termination and to determine the maximum amount of H/OH on the surface, we put 14 H_2O_2 molecules between the ta-C/ta-C interface and let the system relax at 27°C. Then we heated this surface to as high as 1727°C to drive off any weakly bound fragments. This led to 9 (H + OH) terminations on the lower slab and 10 (H + OH) terminations on the upper slab.

7.5c. *Frictional properties of lubricated ta-C surfaces*

Using these surfaces, we carried out sliding simulations by moving the top surface with respect to the bottom one. The ta-C/ta-C sliding couple was constructed by bringing two ta-C slabs into contact. Periodic boundary conditions were imposed in the x-y plane, whereas about 45–48 Å of vacuum was allowed in the z direction. The bottom 0.9 Å of the lower slab (with 32 atoms per cell) were held rigid in all simulations, whereas the top 0.9 Å of the upper slab was slid at a constant velocity of 1 nm/ps along the sliding direction. At each point, all remaining 480 atoms were allowed to move freely according to the forces. An external force was imposed along the z direction (perpendicular to the slabs) to keep the distance between the centers of mass of the two slabs constant. This provides the normal load. Starting with the initial interface models at an initial temperature of 27°C, we carried out constant energy MD simulations while keeping the cell parameters fixed (NVE).

7.5d. *Frictional properties of ta-C surface with glycerol layers*

We studied the reaction of glycerol molecules intercalated between the two H/OH-terminated ta-C surfaces previously described by adding four glycerol monolayers, and each layer is composed of six glycerol molecules. When sliding the H/OH-terminated ta-C friction couples in the presence of glycerol, we were surprised to observe first the disorder in the layer structure and second the decomposition of glycerol and the formation of many water molecules, as shown in Fig. 7.8, which reports the result obtained in presence of four glycerol monolayers. Ethane and glycoaldehyde molecules are also formed in the interfacial region. The number of water molecules

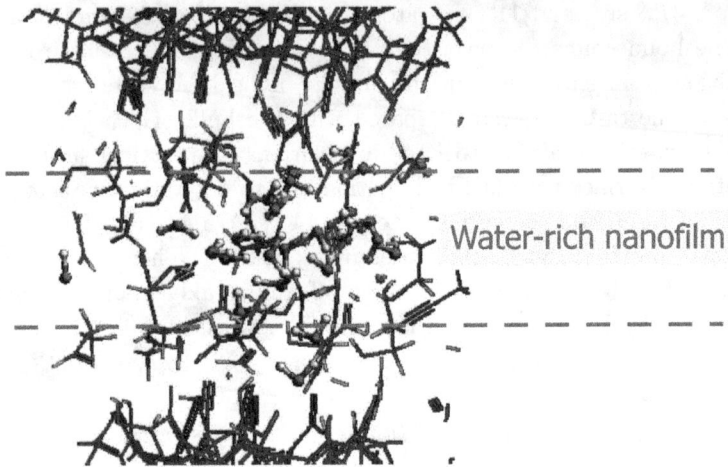

Fig. 7.8. Formation of water molecules by molecular dynamics simulation (shown with small circles) during lubrication of OH-terminated ta-C/ta-C in the presence of initially 4 layered glycerol molecules. Contact pressure is 0.5 GPa and surfaces were slid for 4 ps with a sliding speed of 1 ps.

increases from zero to about 14 after sliding 4 ps. The water molecules, formed by decomposition of glycerol, stay mostly flat on the surface, preventing strong connection with OH groups on the surface by strong hydrogen bonding, and may be sustaining the low friction of the surfaces. The calculation of the temperature in the glycerol layer indicated the possibility of high values at the beginning of sliding and it is probable that glycerol molecules are thermally degraded, according to the simulation. With this assumption, and taking into account the contact pressure of 0.5 GPa, water would be in a supercritical phase ($P > 22.1$ MPa, $T > 374°C$). In the literature, the degradation of glycerol has been explained in a model consisting of two parts: a free radical and an ionic system of reaction pathways. Also, we studied the effect of contact pressure on the friction coefficient and we observe as a general trend that friction decreases as a function of the pressure.

Because the formation of water molecules is activated by the pressure/shear combined effect, we can deduce from these simulations that the formation of water molecules may be an important mechanism for friction reduction by polyhydric alcohols and particularly glycerol.

References

1. J. M. Martin, *et al.*, *J Phys Chem C* 58, 114, 5003 (2010).
2. J. G. Dash, H. Y. Fu and J. S. Wettlaufer, *Rep Prog Phys* 58, 115 (1995).
3. H. Liang, J. M. Martin and T. Le-Mogne, *Acta Mater* 51, 2639 (2003).
4. M. Chen, W. H. Briscoe, S. P. Armes and J. Klein, *Science* 323, 1698 (2009).
5. A. R. Konicek, D. S Grierson, A Gilbert, A. V. Sawyer, A. V Sumant and R. W. Carpick, *Phys Rev Lett* 235502, 1 (2008).
6. Y. Morita, T. Shibata, T. Onodera, M. Koyama, H. Tsuboi, N. Hatakeyama, A. Endou, H. Takaba, M. Kubo, C. A. del Carpio and A. Miyamoto, *Japan J Appl Phys* 4, 3032–3035 (2008).
7. M. Chen, K. Kato and K. Adachi, *Tribol Int* 35(3), 129 (2002).
8. R. S. Gates and S. M. Hsu, *Tribol Lett* 17(3), 399 (2004).
9. F. Ilie, *Tribol Int* 39, 774 (2006).
10. J. M. Martin, C. Donnet, T. Le Mogne and T. Epicier, *Phys Rev Lett B* 48, 10583 (1993).
11. J. M. Martin, M. I. De Barros Bouchet, C. Matta, Q. Zhang, W. A. Goddard, III, S. Okuda and T. Sagawa, *J Phys Chem C* 114, 5003 (2010).
12. D. Dowson and P.Ehret, *Proc Inst Mech Eng* J 213(5), 317 (1999).
13. J. L. Mansot and J. M. Martin, *C R Acad Sc* Paris, 13, 579 (1985).
14. P. Vergne, A. Erdemir and J. M. Martin, (eds.) Superlubricity 429, Elsevier (2007).
15. L. Joly-Pottuz, J. M. Martin, M. I. De Barros Bouchet and M. Belin, *Tribol Lett* 34, 21 (2009).
16. C. Matta, L. Joly-Pottuz, J. M. Martin, M. I. De Barros Bouchet, M. Kano, Qing Zhang and W. A. Godard III, *Phys Rev B* 78, 085436 (2008).
17. W. Habchi, C. Matta, L. Joly-Pottuz, M. I. De Barros, J. M. Martin and P. Vergne, *Tribol Lett*, submitted.
18. N. E. Dorsey, *Properties of Ordinary Water-Substance*, New York, 184 (1940).
19. C. Donnet, T. Le Mogne and J. M. Martin, *Surf Coat Technol* 62, 406 (1993).
20. M. Kano, Y. Yasuda, Y. Okamoto, Y. Mabuchi, T. Hamada, T. Ueno, J. Ye, S. Konishi, S. Takeshima, J. M. Martin, M. I. De Barros Bouchet and T. L. Mogne, *Tribol Lett* 18, 245 (2005).

Chapter 8

Aqueous Lubrication of Ceramics

Mitjan Kalin

Faculty of Mechanical Engineering, University of Ljubljana
Aškerčeva 6, 1000 Ljubljana, Slovenia
mitjan.kalin@fs.uni-lj.si

8.0. Introduction

Much of the research and applications literature in "water lubrication" concerns ceramic materials. In particular, silicon nitride and silicon carbide can provide very low wear and low, even super-low, friction under specific conditions. Alumina can also be a very suitable material for use in water, due to its low-wear properties obtained through the formation of tribochemical layers. Despite generally agreed basic tribochemical and mechanical effects, the detailed mechanisms on the molecular and atomic levels are certainly not agreed upon and not yet clear. In this work we present how remarkably the electrochemical parameters, such as surface charge and the associated pH, zeta-potential or solubility, affect the properties of the surface layer, the physico-chemical and load-carrying properties of the lubricating film, the tribochemical reactions and the tribolayer failure modes of water-lubricated contacts: more than 25 specific properties have been shown or recognized to be potential parameters resulting from surface-charge effects. Under boundary-lubrication conditions, this can result in a significantly different severity of the contact conditions and consequently a great diversity in values of friction and wear, compared to those predicted from conventional considerations of water lubrication. We suggest that a more complete set of tribo-electro-chemical effects should be considered and controlled in the aqueous lubricating film in ceramic systems, and that more refined models that also take into account some of these effects should be developed to enable a more accurate prediction of the contact conditions in future aqueous, boundary-lubricated ceramic tribosystems.

8.1. Oil vs. Water Lubrication Technology

Friction, wear and the related tribological behaviors of contacting moving bodies are mostly dependent on the dynamic physical, chemical and mechanical phenomena that occur at the interfaces of these bodies. Under lubricated conditions, depending on the separation of the contacting bodies as a function of film thickness and surface roughness, different lubrication regimes can be calculated (Eq. 1),[1] and also experimentally illustrated with the Stribeck curve, Fig. 8.1.

$$\lambda = \frac{h_0}{\sqrt{R_{qA}^2 + R_{qB}^2}}, \tag{1}$$

where h_0 is the film thickness and R_q is the measured roughness value (RMS) from the contacting surfaces A and B.

Lambda is typically used to determine the lubrication regimes that are in operation, and describe the ability of the lubricating film to separate the contacting surfaces, which directly influences the friction and wear behavior of a particular lubricated system. Typically, oil-lubrication regimes are determined as follows,[2] by using the lambda parameter: $\lambda \geq 3$

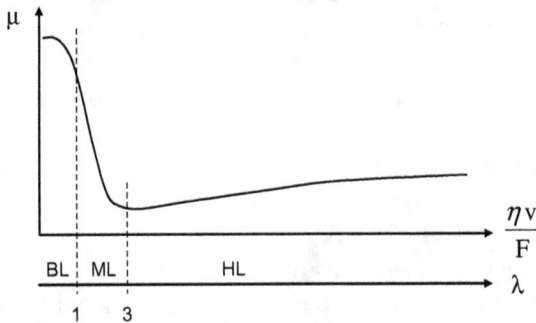

Fig. 8.1. Stribeck curve, i.e. the dependence of friction coefficient, μ, on a dimensionless parameter consisting of the viscosity, η, the speed, v, and the normal load, F, showing different lubrication regimes. The dependence on the lambda parameter is also shown. (BL-boundary lubrication, ML-Mixed lubrication, HL-hydrodynamic lubrication).

for Hydrodynamic, $3 \geq \lambda \geq 1$ for Mixed and $\lambda \leq 1$ for Boundary regime, which is schematically presented in Fig. 8.1.

The boundary-lubrication regime lies at the very left-hand side of the Stribeck curve and is characterized by the dominating influence of the interactions between the contacting bodies, where the hydrodynamic effects of separating the surfaces are of minor importance. Thus, the tribological response of these systems is very complex, since the load and velocity accommodation occur through the many interactions between the contacting asperities, their deformation, tribochemical reactions, the physicochemical phenomena and the resulting lubricating films being of various types, scales and properties, all of which can change significantly, even within the contact, over periods of milliseconds.

In conventional oil lubrication, additives play a key role in this regime, because the surfaces alone are usually not able to provide sufficient protection against adhesion, which results in tearing of the weaker surface, formation of transfer films, subsequent abrasion due to work-hardening of these films, and in the most severe case vibrations, temperature increase and even catastrophic surface stiction. Accordingly, specific additives have been designed to provide relevant protection in oils — depending on contact severity. These systems are designed to be very effective under boundary lubrication even under the most severe conditions. It is of course very challenging to also provide high environmental compatibility at the same time.

Although legislation has been changing over recent decades and becoming more demanding in terms of environmental protection, conventional oils and additives are still focused on function rather than environmental effects. However, ecological awareness and legislation are playing an increasingly important role in supporting replacements of conventional oil systems in the machinery used in forestry, agriculture, mining, off-shore, food processing, etc., i.e. in situations where accidents and oil-spills can have significant impact on environmental pollution. For this purpose, biodegradable oils[3−7] or water-based systems[8,9] can be applied. Therefore, conventional synthetic or mineral oils are now not the only "lubricant" or fluid medium in mechanical systems subjected to tribological conditions.

Additive packages required for boundary-lubrication protection with bio-degradable oils can be quite close or even the same as in conventional oils; however, water is a very different lubricant, both in terms of chemical and physical properties. Accordingly, principles of lubrication design, "additivation" and materials selection are obviously completely different.

However, applications that would use water as a lubricant exclusively for environmental reasons are still rare, which means that the economic market is also very limited. Because of this, efforts and funds to support research in developing tailored additives cannot compare with those dedicated to conventional oil-lubrication and therefore the solutions are at quite an early and under-developed stage. However, green technologies and green tribology is becoming more important and this will certainly bring more research and high-performance results in the field of water lubrication. Of course, water is sometimes employed also for technical reasons or cost. For example, water is a common fluid in many systems that use pumps, turbines, valves, etc., which certainly enlarges the potential use of water-lubrication technologies.

Since water is a poor lubricant due to its low viscosity as well as its poor boundary lubrication properties, friction and wear are always important or even major issues in water-lubricated applications, which are very often dominated by the boundary-lubrication regime. However, water lubrication has typically been managed and addressed mainly through a combination of materials and contact conditions, rather than other more advanced technological solutions, such as additives.

Steel is not a favorable tribological material to be used with aqueous lubricants, due to high friction and corrosion. Polymers are sometimes used due to their corrosion resistance and/or sometimes low-adhesion properties, but polymers are also limited in terms of the applied load and thermal effects, and sometimes also due to machinability and water-absorption issues. On the other hand, various ceramic types have been found to be appropriate materials to overcome most of these disadvantages and are thus often used in water-lubricated applications, such as face seals, taps, etc. Most ceramics provide moderate friction in water, while wear is very favorable for most of the ceramic types due to formation of various wear-protective tribochemical layers.[10–19]

Thus, a lot of the research and applications of "water lubrication" relate to ceramic materials. In particular, silicon nitride and silicon carbide can provide very low wear and low friction under specific conditions.[10,11,20,21] Alumina can also be a very suitable material for use in water due to its low-wear properties; however, the friction is still relatively high to moderate.[14,16,22,23] On the other hand, zirconia is one of the least appropriate types of advanced ceramic for water-lubricated applications because of the (hydrothermal) tetragonal-to-monoclinic transformation[24] and the associated high wear and friction.[25,26]

8.2. Super-Low Friction of Non-Oxide Ceramics

Due to the low viscosity of water, lubricating water-films are extremely thin. Despite this, silicon nitride and silicon carbide can experience, under specific conditions, very low wear and friction. In fact, super-low friction coefficients of 0.002 and 0.0035 have been measured for silicon nitride[10,11] when lubricated with water.

The key mechanisms that are responsible for such excellent properties are not fully clear and are still under investigation. Some researchers claim that it is primarily the tribochemical reactions of ceramics with water that are responsible for the excellent low-friction properties.[27-29] The tribochemistry of silicon nitride and silicon carbide surfaces in water involves the formation of silica. These tribochemical reactions are well known:

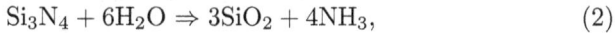

$$Si_3N_4 + 6H_2O \Rightarrow 3SiO_2 + 4NH_3, \tag{2}$$

$$SiC + 2H_2O \Rightarrow SiO_2 + CH_4. \tag{3}$$

Although it has been confirmed that the chemical reaction products from SiC and Si_3N_4 in water are the same,[29] the induction time to achieve super-low friction is different for SiC and Si_3N_4.[30] Figure 8.2 shows that under the same conditions, super-low friction is obtained much faster with silicon nitride than silicon carbide. However, this can be explained by their different kinetic rates, as well as by the presence of hard SiC particles that

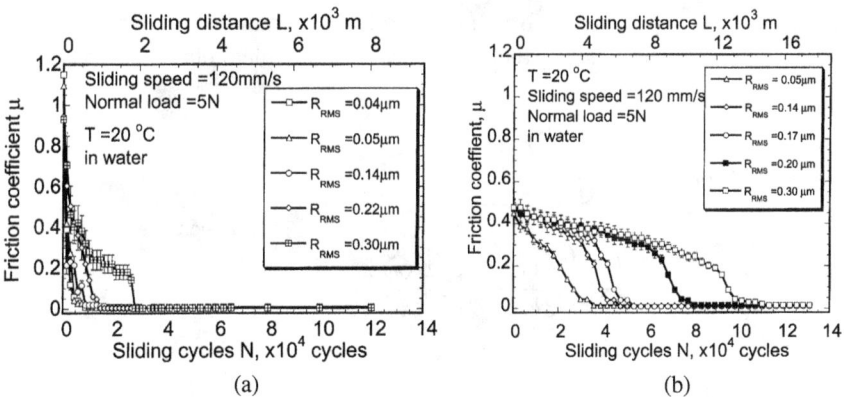

Fig. 8.2. (a) The variation of friction coefficient with sliding cycles of Si_3N_4 under water lubrication. (b) The variation of friction coefficient with sliding cycles of SiC under water lubrication. From Ref. 30.

have been found in the contact and may prolong the running-in of silicon carbide, while this does not seem to be the case with silicon nitride.[29]

However, the reason why silica in water leads to such excellent tribological behavior is not obvious. One explanation proposes that the formation of colloidal silica, leading to a boundary film at the ceramic surface, is the key phenomenon. Two different mechanisms related to the formation of colloidal silica have been suggested. One of the proposed mechanisms for the low friction that is obtained with silica in water is due to the hydrogen bonds formed at the OH-terminated amorphous silica surface.[31] These hydrogen bonds can be easily broken and can lead to a very low shear strength at the surface of the hydrated amorphous silica.[27] The second explanation suggests the important effect of the electrical double layer. Namely, part of the ions that are closer to silica are strongly adsorbed (Stern layer), while those at larger distance have weaker attraction to the surface (diffusion layer).[32] The ions in the diffusion layer can more easily move and thus form a slip plane within the diffusion layer, thus leading to the low-friction behavior. Accordingly, excellent lubrication properties of silica were mainly attributed to slipping of the diffusion layer. The principle of the slipping plane within an electric double layer, which location in reality is, however, not very precisely defined, is schematically shown in Fig. 8.3.

Although the above discussion by Kato and others provides a reasonable tribochemical explanation for low friction, they partially agree with the very different and quite opposite mechanism proposed by Fischer and co-workers. Namely, the latter group proposed that based on tribochemically assisted smoothening of the Si_3N_4 and SiC surface, the newly generated surfaces can be smooth enough to promote hydrodynamic lubrication.[10,20,33] Their argument against a predominantly tribochemical effect was based on

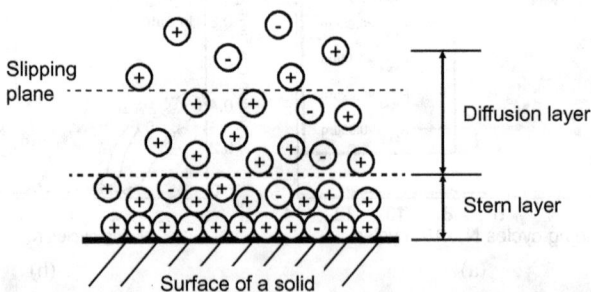

Fig. 8.3.   Schematic model for the low-shear slipping plane within the diffusion layer of the electric double layer.

the Stribeck curve measurements. Namely, if the tribochemistry is the key mechanism, the boundary layer should provide equally low friction at low velocities, but their measurements did not confirm this; on the contrary, the friction rose to value of 0.7 and in-situ FTIR as well as AES did not confirm the required reaction product layer. Moreover, roughness and smoothening of the surfaces are extremely important in hydrodynamic lubrication and the results by Kato and co-workers clearly show the very important effect of these parameters on the super-low friction of these materials,[30] Fig. 8.2. Accordingly, based on several studies, as well as film thickness calculations suggesting water films ranging from 5 nm to 80 nm under various conditions (see Fig. 8.4), this remains one of the most likely responsible mechanisms for the super-low friction behavior.[10,11,20,33]

Nevertheless, some arguments still indicate that the film thickness of these water films may not be sufficient for such excellent friction results. However, if the viscosity of the "water-film" could be higher by an order of magnitude, the suggestion of hydrodynamic or even soft-elastohydrodynamic lubrication would be much more plausible. As mentioned above, the colloidal silica layer could increase the effective viscosity and may be one of the reasons for low friction due to an apparent hydrodynamic effect. Another suggestion that helps in explaining the hydrodynamic

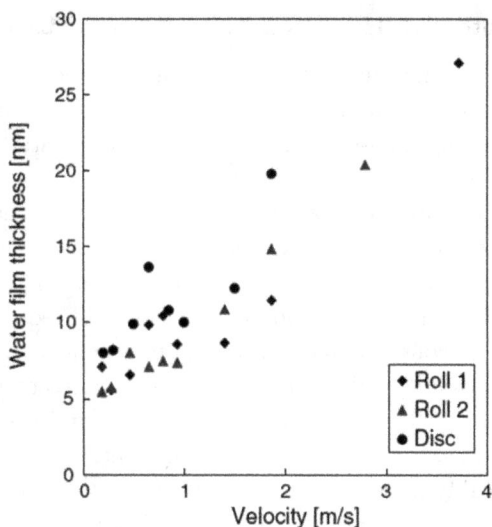

Fig. 8.4.   Estimated thickness of the hydrodynamic fluid film for a silicon nitride bearing operating in water at 2 N load. From Ref. 10.

effects in water lubrication of ceramics is also the formation of a silica gel.[29] Namely, it was proposed that according to reaction:

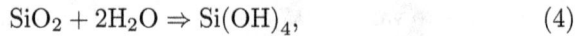

$$SiO_2 + 2H_2O \Rightarrow Si(OH)_4, \qquad (4)$$

silica surfaces and nanoscale silica particles react with water and form silicic acid and silica gel.[29] This gel can enhance the actual viscosity in the vicinity of the surface and thus promote hydrodynamic lubrication.

From above discussion it is obvious that the actual mechanism of the super-low friction of non-oxide ceramics in water is not yet fully clear. Moreover, the proposed mechanisms are quite different: ranging from the extreme left-hand side of the Stribeck curve in the boundary regime, to the mixed lubrication regime, and even to the right-hand side of the Stribeck curve, i.e. in the hydrodynamic lubrication. Despite many studies and thorough theoretical and experimental analyses it seems that some other phenomena may still need to be discovered to fully explain this behavior. Moreover, these studies have also shown that the transition from boundary to hydrodynamic regimes in water is very sharp and could occur only due to small changes (in aqueous solution and/or surface), indicating higher sensitivity to changes in conditions than usually considered in oil lubrication.

8.3. Wear-Protective Hydrated Tribochemical Layers

When an oxide ceramic, in particular alumina, is sliding in water, tribochemical layers are formed. They have the typical appearance of hydrated tribochemical layers, as observed many times in the past;[14-19] an example is shown in Fig. 8.5.[18]

The tribochemical layers appear brittle in nature, as evidenced by many cracks. However, these cracks appear to be a consequence of a drying process of the hydrated layer after the test, not its brittle nature during the tests.[19] Consequently, the properties of the hydrated layers during the experiments are viscous-like and act as a tribochemical "paste", which — on account of tribochemically-dominated wear — protects against mechanical wear of ceramics, which would otherwise occur in a brittle manner and cause high wear. In the case of alumina, these hydrated layers were often non-disputably found to be Al-hydroxide.[16,34] There exists evidence suggesting that wear protective hydroxides (Zr-OH) also form at zirconia surfaces under high humidity,[35] however, the zirconia behavior is strongly dependant on the tetragonal-monoclinic transformation, which may lead to

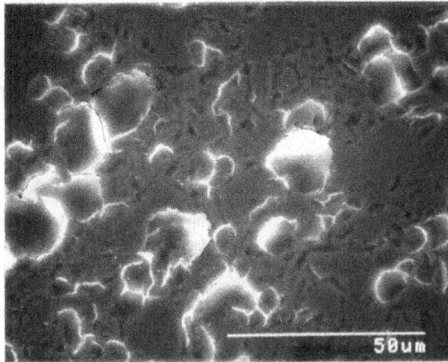

Fig. 8.5. SEM micrograph showing the wear scar with typical hydrated tribochemical layer on an alumina ball after sliding against alumina In-ceram disk (40 N, 2.5 mm/s, 432 m) in distilled water. From Ref. 18.

other undesirable effects. However, the hydrated layers do not result in low friction, but friction is typically moderate-to-high, having values between 0.2 and 0.6,[14,16,36−39] depending on the sliding conditions and specific properties of the layer. A schematic presenting differences in the wear mechanisms of Al_2O_3 ceramic sliding under non-lubricated and water-lubricated conditions are presented in Fig. 8.6.

Despite the more than twenty years of research in this and related fields, some extremely important effects have not been fully recognized and investigated as critical parameters for the wear and friction properties during (boundary) lubrication in water. Until recently, the pH of the solution, which is associated with the surface charge and the zeta-potential (ZP), and which in turn affect the electrochemical and other properties of surfaces (bulk and particles) in aqueous solutions, was not investigated systematically for engineering contacts in order to analyze its effect on the wear mechanisms and tribological behavior in various regimes. However, in our laboratory we have found that the pH, which influences the surface charge and ZP in aqueous solutions, affects the wear of alumina and zirconia by an order of magnitude, and the friction by a factor of 2-3.[26,36,37,40] Therefore, the variations in the tribological performance observed within the boundary-lubrication regime for different pH values of an aqueous solution, could be even larger than the typical changes of these properties between different lubrication regimes, i.e., boundary-mixed-hydrodynamic.

Figure 8.7 shows — schematically — how the surface charge is developed under different pH and ZP conditions. In preferentially acid

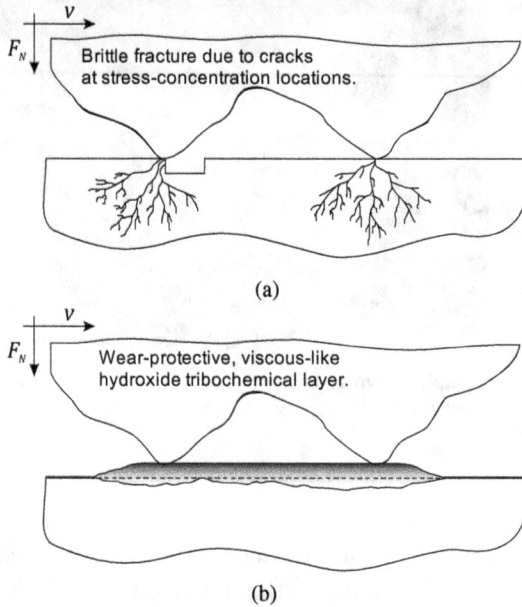

Fig. 8.6. Schematic presentation of the wear mechanisms of oxide ceramics sliding under: (a) non-lubricated conditions; no tribochemical film, brittle fracture with cracks due to overloading, preferentially at stress concentration locations, and (b) water-lubricated conditions; hydroxide tribochemical layer prevents brittle fracture.

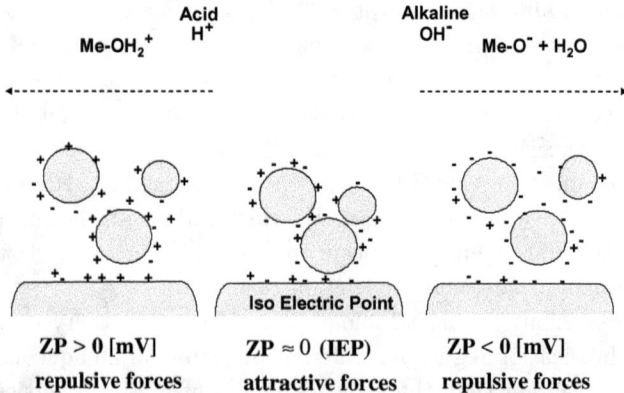

Fig. 8.7. Schematic presentation of the electrochemical mechanism of pH and the electric charge at the surfaces (bulk, wear debris) in water. From Ref. 39.

environment, the surfaces and particles will exhibit positive ZP, with mainly positive ions at the surface, while in preferentially alkaline environment, ZP will be negative, with mainly negative ions surrounding the surfaces (including particle surfaces). In contrast, at iso-electric-point (IEP), the net surface charge of positive and negative ions will be equal, thus ZP will be zero.

With this phenomenon in mind, we tentatively suggest some new influencing parameters and a new tribo-electro-chemical concept for the boundary-lubrication evaluation that should be developed and taken into account in order to meet the requirements of boundary-lubricated systems in the future.

Moreover, large differences in the proposed mechanisms for low-friction behavior in non-oxide ceramics, as discussed extensively in Sec. 2, also confirm the necessity to consider new, significant parameters. Furthermore, often referred model for improved tribological properties due to electric double layer associated with formation of silica in SiC and Si_3N_4 systems in water is not completely clear, because it assumes that silica particles are positively charged. However, for silica it was reported by many authors, and is summarized in Ref. 41 that the IEP lies at a pH between 1.8 and 3 (Fig. 8.8). Accordingly, the ZP of silica at a pH that is typical in most macroscopic aqueous tribological experiments (around pH 6 or 7) should be considered negative, which may change the proposed mechanisms and effects — which would certainly depend on pH and ZP. However, as described above, the surface charge, ZP, and other associated electrochemical phenomena, including dissolution and formation of various ions, all

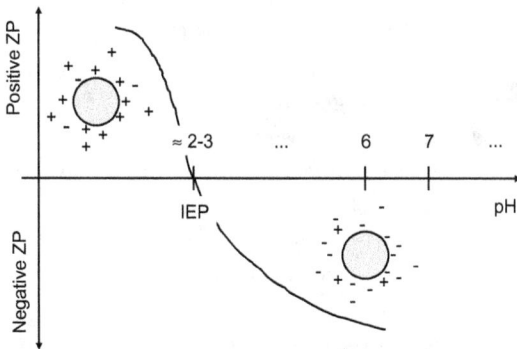

Fig. 8.8. Schematic presentation of dependence of ZP and IEP on pH for silica, based on data in Ref. 41.

depend on pH value, and are all clearly inter-dependant. Therefore, the actual electrochemical behavior could be very complex, and could influence tribological behavior in different ways — depending on ZP and pH.

8.4. The Electrochemical Mechanism of pH and the Electric Charge at the Surfaces in Water

Despite the lack of thorough investigations of the influence of the surface charge and ZP in macroscopic tribological contacts, the electrochemical effect has been recognized in the past as playing a role in water lubrication.[11,27-29,42-44] Moreover, in earlier studies, it was suggested that a double layer that forms at charged surfaces in electrolyte solutions could cause an electro-kinetic repulsive force,[45,46] which serves to reduce friction in contacts of silicon nitride and silicon carbide in aqueous solutions.[11] Namely, a streaming current in the gap of two charged bodies sliding over each-other alters the Maxwell stress field[45,46] in the gap and forms an additional normal repulsive force that helps to separate the two bodies. This is similar to what has been suggested for alumina, proposing such repulsive action due to electrochemical effects in aqueous solution.[42] However, only the repulsive force was considered in these studies.

On the other hand, frictional forces, dissipated energy and adhesion on a nanometer scale were studied using LFM on silicon nitride, silica and titanium dioxide in aqueous solutions in the range of pH 3 to pH 9.[43,44,47] Although study[47] suggested that friction is independent on pH, other two studies[43,44] supported effect of pH and surface charge on all these parameters, both in the region of attractive and repulsive interactions. It is interesting to note that lateral force and dissipated energy results well resembled the characteristic "S" shape of the zeta potential curve (as a function of pH). This indicated the significant effect of surface charge and ZP on both normal and tangential forces at AFM tip. Accordingly, by applying low loads, which were close to those of the electrostatic forces (in the range of nN), these experiments demonstrated the importance of surface charge, ZP and pH on the nanoscale friction and adhesion. The rapid decay of the lateral force through the iso-electric-point (IEP)[43,44] can be explained by the steep slope of the zeta potential curve in the region close to the IEP.[39]

However, the effects of the electrochemical conditions close to the IEP are more profound and cause more surface effects than just the

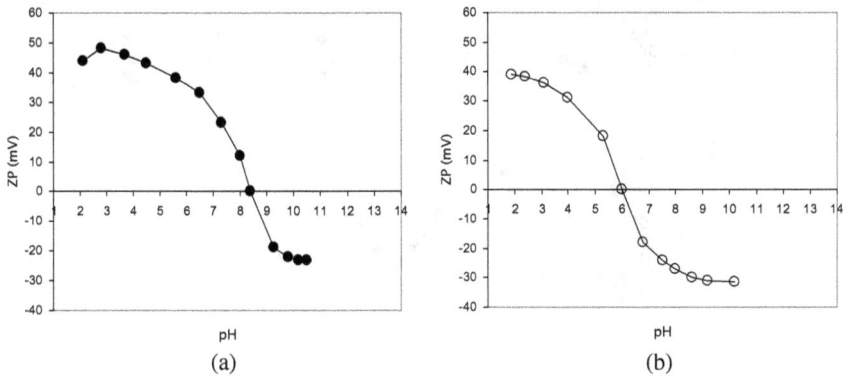

Fig. 8.9. Zeta-potential as a function of pH for (a) alumina, and (b) zirconia.

addition/ subtraction from the applied external load, especially at the macroscopic-scale contacts. Evidence from experiments at much higher external loads compared to the values of these electric forces shows that the effect of the electric surface charge on the tribological behavior is still noticeable, even though the electric forces can be virtually neglected due to their orders-of-magnitude lower size.[36,37,40,47]

The IEP (zero net surface charge) is defined by the value of the zeta-potential (ZP) being equal to zero for this condition.[48] Moreover, there also exists a well-defined correlation between the ZP and the pH, which is characteristic for a particular material, as seen in Fig. 8.9 (and schematically in Fig. 8.8). Therefore, for a specific material a relationship between the IEP and the pH can also be established.

ZP measurements of the surfaces in engineering systems are rather difficult to perform and thus inconvenient for everyday use in engineering practice and water solution control. In contrast to the ZP, however, the pH is a rather simple, inexpensive and broadly accessible parameter to measure and control. Moreover, for a particular system, it is even possible to modify their relationship by using different polyelectrolytes and so shift the ZP curve with respect to pH value, thus obtaining, for example, the IEP at a different pH value than the characteristic one for the unmodified oxide, see Fig. 8.10. This gives us a lot of possibilities to control the surface charge across a wide range of aqueous solutions by simply controlling the pH value, and even maintaining the same surface charge at different pH values, which are extremely important properties in engineering practice.

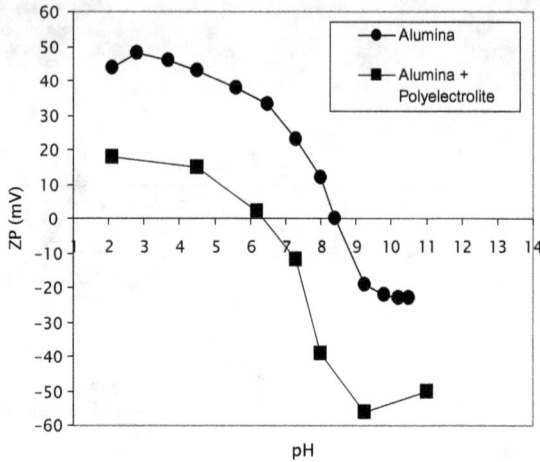

Fig. 8.10.   Shift of zeta-potential curve with respect to pH when using a polyelectrolyte (Dolapix CE 64, Zschimmer&Schwarz, Lahnstein, Germany). From Ref. 39.

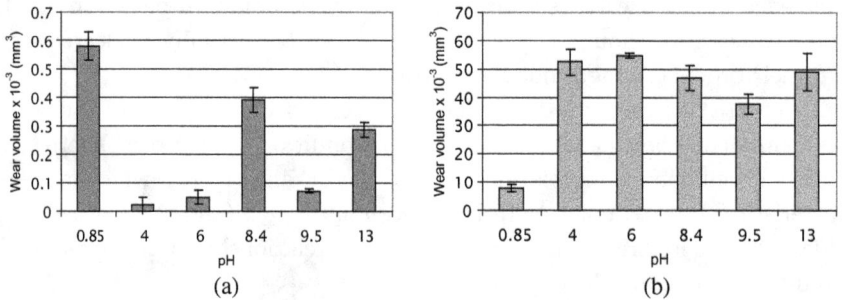

Fig. 8.11.   Wear volume after 100 m of sliding at 50 N as a function of pH of (a) alumina, and (b) zirconia pins. Note that the scale in Fig. (b) is 10-times larger. From Refs. 26,36.

8.5. Effects of pH and Surface Charge on Tribological Behaviour and Formation of Boundary Surface Layers of Oxide Ceramics

8.5.1. *Wear and friction behavior*

Figure 8.11 shows the results of the experiments in self-mated contacts for alumina and zirconia across a broad range of solutions with different values of pH. A big difference (one order of magnitude) in wear in a very narrow region around pH 8 was observed for the alumina. This shows a high sensitivity of alumina in terms of wear-resistance in the pH region

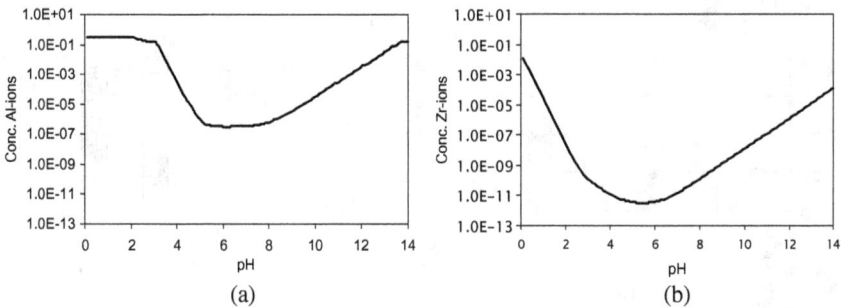

Fig. 8.12. Solubility of (a) alumina and (b) zirconia in water solutions as a function of pH. From Ref. 39.

that is close to solutions typical for most practical applications, i.e., tap or distilled water, which commonly have a pH of 6 to 8. The wear-maxima in this application-relevant pH region, i.e., at pH of approximately 6, were also found for zirconia, but the effect is less pronounced. However, it should be noticed that due to the much higher wear compared to alumina, these absolute differences in wear loss are even higher. A comparison with the pH value at which the IEP occurs for alumina or zirconia (see Fig. 8.9) reveals that the wear maxima exactly coincide with this electro-chemical parameter, which represents the conditions of zero net charge at the surface. However, significant differences in the wear of both materials are observed in very acidic environments (pH < 1), and for alumina also under alkaline (pH > 13) conditions. In these extreme cases, the wear is associated with the solubility of the alumina and zirconia in aqueous solutions, Fig. 8.12. As seen, the highest values are observed at the extreme left-hand and right-hand sides of pH scale, i.e. in very acid or alkaline environments.

The coefficient of friction (Fig. 8.13) is also very dependent on the solubility, being clearly lower in the regions where the solubility is high, Fig. 8.12. However, in the intermediate pH range, where surface charge plays a dominant role ($\approx$ pH 4 − 10), the coefficient of friction is similar and relatively high; higher than in the solubility-determining pH region.

8.5.2. *Thickness of the tribochemical layer*

The appearance of the worn surfaces that are formed in different aqueous solutions depends significantly on the pH value and the electrochemical properties of the aqueous solution, i.e., surface charge and solubility, as could be anticipated already from the empirical wear and friction results in

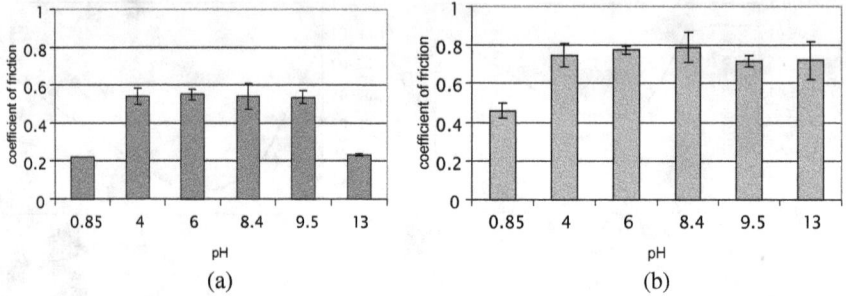

Fig. 8.13.   Steady-state coefficient of friction as a function of pH in contacts of (a) alumina, (b) zirconia. From Refs. 26,36.

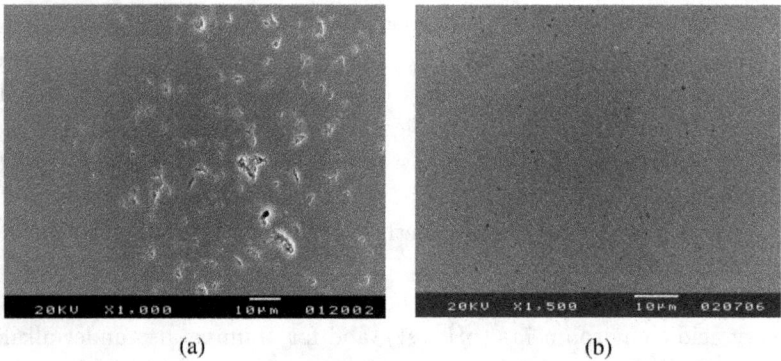

Fig. 8.14.   SEM images of worn surfaces in the pH region with the predominant solubility effect of (a) alumina at pH = 0.85, and (b) zirconia at pH = 0.9. From Refs. 26,36.

Figs. 8.11 and 8.13. From the SEM analyses it is clear that solubility plays a critical role in the very acidic and alkaline pH regions, i.e., under conditions where its effect is also expected to be the largest, Fig. 8.12. Namely, under these conditions, no wear debris or any other surface features (except some pits on the alumina surface) can be observed at the wear scars, suggesting a dominating influence of the dissolution of the materials, Fig. 8.14.

On the other hand, in all other pH regions, there are always tribolayers formed on the wear surfaces. However, the properties of these layers strongly depend on the pH value and the associated surface charge under those conditions. It is clear that close to the IEP the tribolayers are very thick and consist of heavily compacted wear debris, which subsequently fractures and delaminates, Figs. 8.15(a) and 8.15(b).

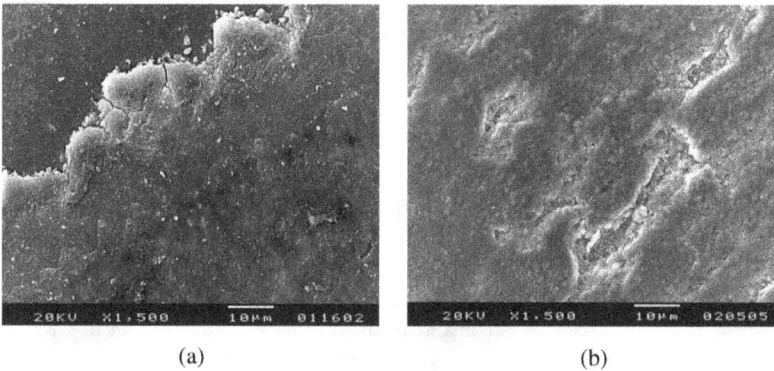

Fig. 8.15. SEM images of thick tribofilms formed in the pH region at IEP or close to IEP of (a) alumina at pH 8.4 (IEP), and (b) zirconia at pH 6 (IEP). From Refs. 26,36.

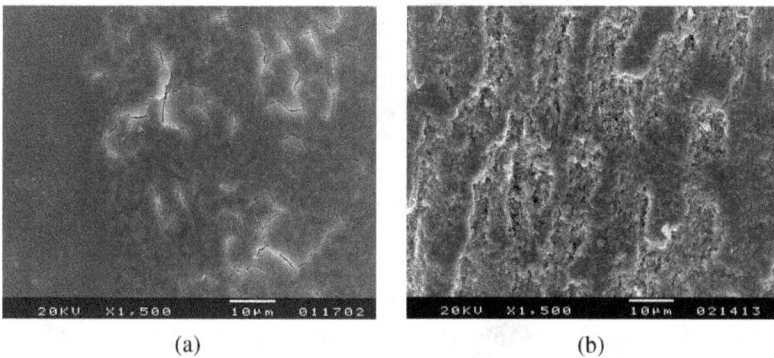

Fig. 8.16. SEM images of thin and/or non-continuous tribofilms formed in the pH region with large surface charge (away from IEP) of (a) alumina at pH 4, and (b) zirconia at pH 3.9. From Refs. 26,36.

In other pH regions where a tribolayer is also formed, but the surface charge is high (away from the IEP), the layers become thinner and less consistent (Figs. 8.16(a) and 8.16(b)), which coincides with the larger electrostatic repulsive forces between the surfaces and the debris (compare Figs. 8.7 and 8.9), which tend to keep them separated.

Figure 8.17 presents the profilometric measurements of the tribolayers under various surface-charge conditions. It can be seen that the thickness of the layers at the IEP is indeed rather high: values between 1 and 2 μm can be estimated for the ceramics used, Figs. 8.17(a) and 8.17(b). And, in agreement with the SEM observations, as the ZP and surface charge

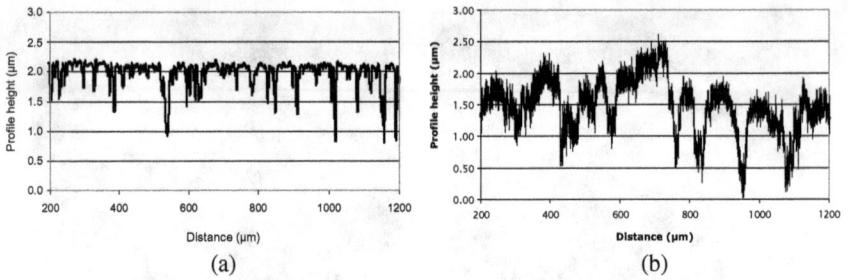

Fig. 8.17. Profilometric measurements of thick tribofilms of (a) alumina at pH 8.5 (thickness $\leq 1.5\,\mu$m), and (b) zirconia at pH 6 (thickness $> 2\,\mu$m). From Ref. 39.

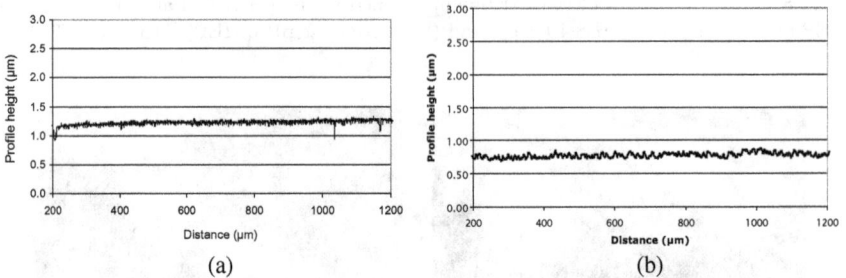

Fig. 8.18. Profilometric measurements of worn surfaces that do not form tribofilms. (a) Alumina at pH 0.85 (no layer), and (b) zirconia at pH 0.9 (no layer). From Ref. 39.

increase from the IEP toward higher negative or positive values, the layers become thinner, until finally, in the pH region with the predominant solubility effect, they completely disappear, resulting in very uniform topographical properties, Figs. 8.18(a) and 8.18(b).

8.5.3. Roughness and texture of the tribochemical layer

Because of the gradual thinning of the layer, the texture, which is characterized by the surface un-evenness, waviness, valleys, peaks, pits and other effects, is also changing. To quantitatively evaluate these changes, roughness can be used in the first approximation, as the most standardized and simple parameter, which broadly reflects the phenomena discussed. Accordingly, from Fig. 8.19 it is clear that the difference in the Rq roughness of the tribolayers is as high as about $0.15\,\mu$m, while for the no-layer conditions, where the solubility effect dominates, the difference is even higher, about $0.25\,\mu$m. If this variation in the surface-asperity heights, which depends

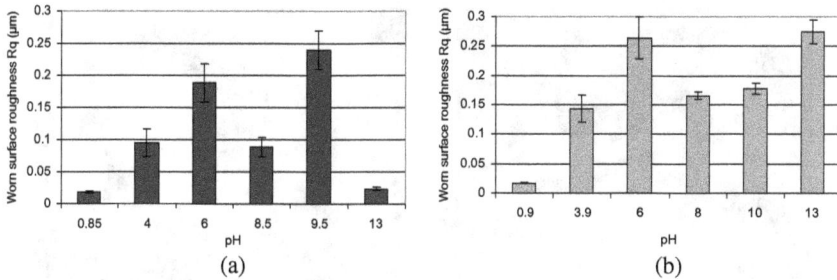

Fig. 8.19. Roughness (R_q) measurements of worn surfaces as a function of pH for (a) alumina and (b) zirconia. From Ref. 39.

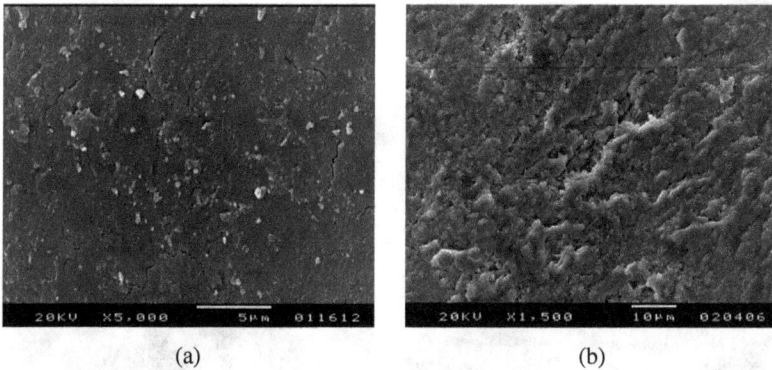

Fig. 8.20. Thick tribochemical layers at IEP show high consistency in morphology and texture: (a) alumina at pH 8.4 (IEP), and (b) zirconia at pH 6 (IEP).

on the pH of the aqueous solution, is compared to the typical calculated thickness of the water film, reported to be from 0.005 to 0.080 μm,[10,27] it is clear that the effect of the surface charge and the pH could be even 10–100 times larger than the thickness of the water film itself.

Moreover, Figs. 8.20 and 8.21 clearly show that the texture and the consistency of the layer are also affected by the surface charge. Namely, it can be seen that close to the IEP the layers are compacted and coherently embedded into a consistent surface layer, obviously affected by the attractive forces, Figs. 8.20(a) and 8.20(b). The surface layer of zirconia (Fig. 20(b)) is, however, fractured due to the specific wear mechanism discussed later, but the consistency and compaction of the layer are still obvious.

On the other hand, tribolayers formed in aqueous solutions where the surface charge is high, i.e., far from the IEP, are formed from small

(a)   (b)

Fig. 8.21. Tribochemical layers formed at non-IEP conditions, show less compacted morphology and texture: (a) alumina at pH 9.5, and (b) zirconia at pH 10. From Refs. 36,39.

(a)   (b)

Fig. 8.22. SEM images of typical thick and compacted wear debris as a function of pH of (a) alumina at pH 8.4 (IEP), and (b) zirconia at pH 6 (IEP). From Ref. 39.

agglomerates or even separate sub-micron-sized wear debris (Figs. 8.21(a) and 8.21(b)) because of the repulsive forces between the surfaces and/or debris, which tend to keep them separated.

8.5.4. Wear-debris generation

In the vicinity of the IEP the wear debris were large and compact. Fig. 8.22(a) shows about $20\,\mu$m large flake of alumina surface. In agreement to wear data, wear debris removed from zirconia surfaces were even larger. Fig. 8.22(b) shows only a part of a dense and very large piece of zirconia

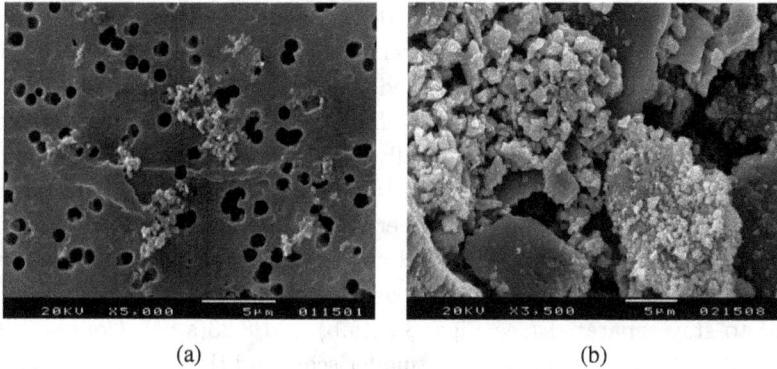

(a)   (b)

Fig. 8.23.   SEM images of wear debris generated from tribofilms at non-IEP pH values of (a) alumina at pH 4, and (b) zirconia at pH 3.9. From Ref. 40.

debris (note that magnifications of Figs. 8.22 and 8.23 are the same), with some smaller fragmented debris on top of it. It is clear that these wear debris were removed from the thick and consistent/dense surface layers, as shown in Figs. 8.15(a,b) and 8.20(a,b). A difference in the size of the debris between alumina and zirconia, which reflects in the wear mechanisms and the wear rate, being much higher and more severe for zirconia (see Fig. 8.11), is also clearly visible.

On the other hand, in the pH region away from the IEP, the wear debris were usually very small (μm-range) and non-compacted, typically fragmented into groups of small separated debris, Fig. 8.22(a) and 8.22(b). This again reflects the effects of the repulsive forces under these conditions and agrees with all the other observations. In pH regions where the dissolution of material dominates (very acidic and/or very alkaline) there are hardly any wear debris observed.

8.5.5. *Wear mechanisms*

Wear mechanisms depend significantly on the pH value and the associated characteristic surface charge or solubility under that condition, but they also depend on the type of material with its specific properties and generic failure modes. For both materials studied there are pH regions where tribolayers never form. These regions are dominated by material dissolution, which is found for zirconia under very acidic conditions (pH<1), and for alumina under both very acidic (pH<1) and alkaline conditions (pH>13), see Fig. 8.12. In the whole intermediate pH region, however, tribolayers are formed. These consist of loose wear debris, which are compacted under a normal load to form layers of different consistency. Close to the IEP the layers are very uniform and compacted, and where they are not spalled they are also smooth. This is due to the effect of the attractive forces between the surfaces and the debris. Typically, the hydroxides that are formed in aqueous solutions also help to embed the wear debris, which become very clear at or near the IEP point. However, as the pH moves from the attractive to the repulsive force regime (the pH moves away from the IEP), the layers become less coherent and the debris are not strongly bonded anymore, but tend to stay separated (see Figs. 8.21(a,b) and 8.23(a,b)). Consequently, the material removal occurs on a smaller scale and the overall wear is also smaller, Fig. 8.11.

To verify and prove the effect of the surface charge, a commercially avalable anionic polyelectrolyte (Dolapix CE 64, Zschimmer & Schwarz, Germany), which is based on a carboxylic acid, was used in aqueous solutions. A more profound study on its effects is presented also in Ref. 49. This shifts the zeta-potential curve with respect to the pH (in the case of alumina from pH 8.4 to pH 6), but without changing the shape of the ZP-pH curve (see Fig. 8.10). The wear behavior under such conditions exactly followed the explained mechanisms and surface phenomena, i.e., the wear-maxima of the alumina shifted from pH 8.5 to pH 6, where the new IEP was established, Fig. 8.24.

It should be noted that the wear of zirconia is about 10–100 times higher than that of alumina (see Fig. 8.11), which is a consequence of

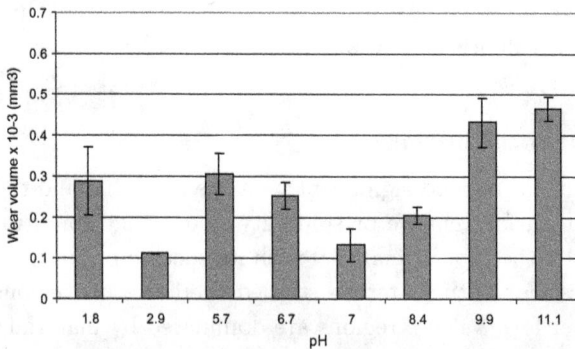

Fig. 8.24.   Wear volume after 100 m of sliding at 50 N as a function of the pH of alumina with the addition of polyelectrolyte (Dolapix CE 64, Zschimmer & Schwarz, Germany) in an aqueous solution.

two other characteristic phenomena, one being wear-promoting for zirconia, the other being wear-protective for alumina. Namely, zirconia is subjected to a tetragonal-to-monoclinic transformation under tribological contact in water, which is associated with a volume increase and substantial surface fracturing, as was observed in the past[50,51] and also in our detailed TEM and XRD studies.[26,40] On the other hand, alumina is known as a suitable material for water lubrication, primarily due to the generation of hydroxides that form soft tribochemical layers. These layers were extensively discussed and characterized in the past, and their beneficial low-friction and wear-protective tribochemical effects are well known nowadays.[14,18,22,34,52] In the pH region where tribochemical layers form, the mechanical wear (fracture and deformation) dominates the wear rate of zirconia, while the soft hydroxide layers with the mild tribochemical wear dominate the wear rate of alumina. It is therefore reasonable that the effect of the electrochemical forces and the accompanying surface effects discussed above are less pronounced in the case of zirconia because of the prevailing mechanical wear mechanism, compared to the mild tribochemical type of wear characteristic for alumina under the present conditions. However, it seems reasonable that the contact severity and the material inherent failure mechanisms will play a role in these co-acting and interdependent mechanisms, which will determine the magnitude of their effect and thus the actual overall wear and friction behavior.

8.6. Comments on Various Influencing Parameters

8.6.1. *Lubricant film thickness and lubrication regimes*

Due to the low viscosity of water, lubricating water films are very thin, thus the "height" of the surface features is close to the film thickness, which means that a small change (nm-range) in the surface and/or film properties makes a big difference in terms of the number and severity of the asperity contacts, i.e. type of lubrication regime, and consequently in the wear and friction behavior. That is to say, even very small variations in surface features, for example, affected by the pH and surface charge, as observed in our work, can cause under such low-thickness and fragile film conditions a change from hydrodynamic to mixed or boundary regime — and vice versa. This is also clear from the very different proposed mechanisms for super-low friction of non-oxide ceramics (claiming both boundary[27−29] and hydrodynamic[10,11,20,33] lubrication).

However, the present state of the art for determining/calculating the lubricating film thickness and consequently the lubrication regime do not consider electrochemical effects such as pH, surface charge or solubility. Currently, the quality of the lubricating film and, indirectly, the contact severity, which determine the friction and wear properties, can be calculated using several film-thickness equations, such as Eq. (5),[2] and the lambda parameter Eq. (1),[1] which indicate the lubrication regime:

$$\frac{h_0}{R'} = 3.63\, U^{0.68}\, G^{0.49}\, W^{-0.073} \left(1 - e^{-0.68\,k}\right), \tag{5}$$

where h_0 is the minimum film thickness, R' is the equivalent contact radius, U is the parameter of the velocity depending on the velocity and viscosity, G is the parameter of material properties, W is the parameter of the load, and k is a parameter that depends on the contact shape (ellipticity).

Clearly, all the influencing parameters considered in the above models are basic/general physical, macro-geometrical (contact/element shape) or mechanical. Moreover, the most significant among all these parameters that currently affect the prediction of the film thickness and the lubrication regime for the selected tribosystem (shape, contact materials, etc.) and the loading conditions (velocity, normal force, etc.) are the surface roughness and the lubricant viscosity.

However, in the present review chapter we have shown that the electrochemical parameters, such as the pH and the surface charge, significantly change the contact surface's appearance with its many characteristic features. Typically, a new tribo-electro-chemical surface film is formed, which has a different thickness (Figs. 8.17 and 8.18), roughness (Fig. 8.19), and texture (Figs. 8.20 and 8.21), with all these directly affecting the parameters in the above calculations. So, by keeping all the loading, material and geometrical parameters the same, but not considering the surface-charge effects appropriately, the calculated values can be significantly different from the actual contact conditions. Even distilled, de-mineralized or tap water can have noticeably different pH values, especially when used from different sources (producer, geographic location, etc.), so the electric charge at the surfaces will also be different, and because of the large changes, and even in a very narrow, application-relevant region (Fig. 8.9), these differences can be substantial. Moreover, the surface charge is affected by polyelectrolytes (additives) and other ions in a specific material/system (see Fig. 8.10), which also affect the lubricating film, and consequently the tribological behavior, Fig. 8.24. Therefore, it is clear that a more detailed definition of the electrochemical parameters, such as pH, ZP and surface charge, is necessary for water-based lubricating systems, in order to improve agreement with the actual contact conditions and also to increase reproducibility and provide a more relevant comparison of the tribological data.

However, in the boundary lubrication regime, this situation is even more complex due to other effects that are difficult to model. In addition to the well-accepted influence of tribochemistry, the texture, morphology and mechanical properties of the tribochemical layers can be very different, but there is little consideration of these effects in present-day management of water-lubricated contacts. Some of these are further discussed in the next section.

8.6.2. *Surface properties*

Typical water-lubrication film thicknesses were reported to be in the range of 0.005-0.08 μm,[10,27] which is rather low compared to the roughness of engineering surfaces. This suggests that extremely smooth and uniformly flat surfaces are required[10] to achieve a hydrodynamic or satisfactory mixed lubrication under water-lubrication conditions. However, the roughness variation between the tribolayers that have formed under different surface-charge conditions (pH) in our work was as high as 0.15 μm, while compared to the no-layer conditions (solubility effect), the difference was up to 0.25 μm, Figs. 8.17 and 8.19. Therefore, the actual (pH-dependent) surface roughness can be 10–100 times higher than the expected water-film thickness or initially calculated "dry" roughness. This further suggests that the actual contact conditions can be completely changed and be a long way from those encountered/calculated in present-day models if the surface charge (pH) is changing.

In addition, not only the roughness, as the most common quantitative parameter, but also the texture, thickness and consistency of the surfaces are greatly affected and changed due to a variation of the surface charge, as seen in Figs. 8.15–8.21. That is to say, when the water-lubrication film thicknesses of 0.005-0.08 μm[10,27] are compared to the variations in the surface tribolayer "heights" and "flatness" due to differences in its thickness, i.e. up to 2 μm in our study (see Figs. 8.17(a) and 8.17(b)), the importance of the surface charge and the pH effect becomes clear. Accordingly, the surface unevenness, waviness, the amount and depth of the valleys and pits (and height of the peaks) that are differently distributed and generated as a consequence of a specific (pH-dependent) tribolayer's surface properties will also affect the lubricating film and the tribological behavior.

8.6.3. *In-situ surface texture and the lubricating film's load-carrying properties*

The valleys and pits of about 1–2 μm thickness (Fig. 8.17) that form due to different electro-chemical conditions and act as in-situ "texturing" in these contacts, will also affect the intrinsic hydrodynamic (load-carrying) properties of the lubricating film. The significant effect of surface texturing on the hydrodynamic properties of the lubricating film was already confirmed by purposely introduced pits of various shapes, surface ratios and depths, including those of the same/similar range as in our experiments, i.e., 4–8 μm.[53–55] Moreover, this technology is nowadays widely used in applications of water-lubricated ceramic systems,[53,54] i.e., in mechanical seals. The effect on lubrication is still under investigation; where hydrodynamic lift due to the pits (Rayleigh bearing effect) and the "lubricant" reservoir are two major possible contributions. This suggests that similar surface phenomena that were observed due to surface-charge variations, are of great importance and may have a significant impact on the load-carrying capacity of the thin water-based lubricating film's properties. Accordingly, this supports the idea that these surface effects should be carefully considered in boundary (or mixed) lubricated fluid films or even incorporated into future film-thickness and lubrication-regime models.

8.6.4. *Mechanical properties and failure mode of the tribochemical layer*

Another important consequence of surface charge due to repulsion and/or attraction between the surfaces and debris is the effect on the mechanical properties of the newly formed electro-tribochemical layers. These properties have not yet been directly measured, because much more effort needs to be made due to their extreme brittleness after drying. However, from our surface analyses (see Figs. 8.14-8.23) some effects seem to be clear. Namely, the tribo-layers that are formed in the pH regions away from the IEP consist mainly of small agglomerates or even separate sub-micron-sized wear debris (see (a) and (b) in Figs. 8.16, 8.21 and 8.23) because of the repulsive forces between the surfaces and/or debris, which tend to keep them separated. Such layers are certainly less robust and most probably have different mechanical properties (hardness, elasticity) to the layers at, and close to, the IEP, where attractive forces keep the debris and layers together, (see (a) and (b) in Figs. 8.15, 8.20 and 8.22). Consequently, this affects the deformation behavior, which further influences the real contact area and the contact stresses, and finally the wear mechanisms under tribological conditions.

The wear performance and wear mechanisms that were presented here certainly support the above discussion. Namely, if the mechanical properties of the layers are different, their response to external loading will most probably result in a different failure mode, too. In our work, the wear rates (Fig. 8.11) and the appearance and amount of wear debris (Figs. 8.22 and 8.23) and surface layers (Figs. 8.15–8.18, 8.20 and 8.21) consistently show that the failure modes of the layers under different pH conditions were indeed significantly different. In agreement with the discussion so far, the material-removal mechanism for weakly bonded small-sized debris layers dominated by the repulsive forces was small-scale "single-debris" removal and the wear and the debris were, as a consequence, small. On the other hand, strongly and coherently bonded layers due to attractive forces at the IEP experienced the delamination and spalling of large flakes (Figs. 8.22(a) and 8.22(b)), which led to high wear, Fig. 8.11.

8.6.5. *Viscosity and viscosity-related properties of the lubricating film*

Although the phenomena observed and discussed in this work are not restricted only to lubrication with water-based solutions, it seems that the effect of viscosity variation as a function of surface charge (pH and ZP) on the lubricating film's properties is more relevant for (low-viscosity) water than for conventional oils. It is well known that the viscosity of aqueous solutions could, depending on the material and the suspension concentration, change by orders of magnitude,[48] as presented in Fig. 8.25. It is therefore clear that the lubricating film's properties will also be affected by this important and independent parameter, which directly affects the lubricating media, not the contacting surfaces. Moreover, as mentioned previously, the viscosity of the lubricating film is recognized (for conventional lubricants and models) as one of the most influential parameters for the film thickness and lubrication regime. In contrast to the negative influence of the (increased) roughness that was found in our work at the IEP, the viscosity of aqueous solutions increases at the IEP (see Fig. 8.25) and, therefore, it is reasonable to expect it to have a beneficial effect on the lubricating film and

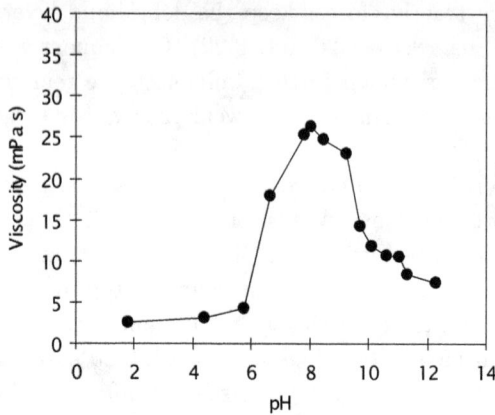

Fig. 8.25. Viscosity dependence on pH (surface charge) for a 40 at. % suspension of alumina in water. From Ref. 39).

the tribological performance. This is the opposite to the tribological behavior observed in our work; however, the effect of viscosity was not "isolated" and investigated in detail, so it is not possible to verify its actual influence. Therefore, the effect of the viscosity change of the aqueous solution at different pH values on the tribological properties and its dependence on pressure, temperature, shear, etc., i.e., the characteristics that are relevant for tribological properties, remain completely unknown at present.

Actually, this effect was already suggested for non-oxide ceramics due to the previously discussed excellent super-low friction properties and associated potential increase in effective viscosity. Silica-gel and colloidal silica layers were proposed to explain this behavior.

8.6.6. *Final remarks*

There are other parameters that affect the overall behavior and properties of water-lubricating films that have not been discussed in detail in this chapter. That is to say, the type of materials and tribochemical reactions will also play a significant role in the wear mechanisms, as already recognized previously, and this will certainly affect the boundary-lubrication film properties, but also the interface shear strength. Some of the effects of the tribochemical reactions on the interfacial shear and friction, which are closely related to the electro-chemical properties, have already been recognized on different scales and were presented in Refs. 11,27,42 and 43 An additional influence is the solubility of materials under different pH conditions, which is obviously very relevant for ceramics and water solutions.[36,37,40] However, as we have found, these have an effect only in very acidic or alkaline conditions. Therefore, they might be less relevant for practical applications than the surface charge, which is more effective in the moderate pH region, where most engineering systems operate. Nevertheless, because the ZP curve can be shifted to different pH values (see Fig. 8.10) by using various polyelectrolytes,[48] any region and the inter-dependence of the solubility with the surface charge should also be considered in the future. Many of the recognized (shown/proven and potential) effects of surface charge and the associated pH and ZP on the in-situ surface properties and/or lubricating film quality discussed in this work are shown in Fig. 8.26.

Of course, it will probably not be possible to incorporate all the parameters that influence and are influenced by the variation of the electrochemical properties into one or even a few computational models. However, the examples and/or potential effects shown here justify the need for an awareness of these processes and a suggestion that many of these effects should be carefully considered when investigating and analyzing

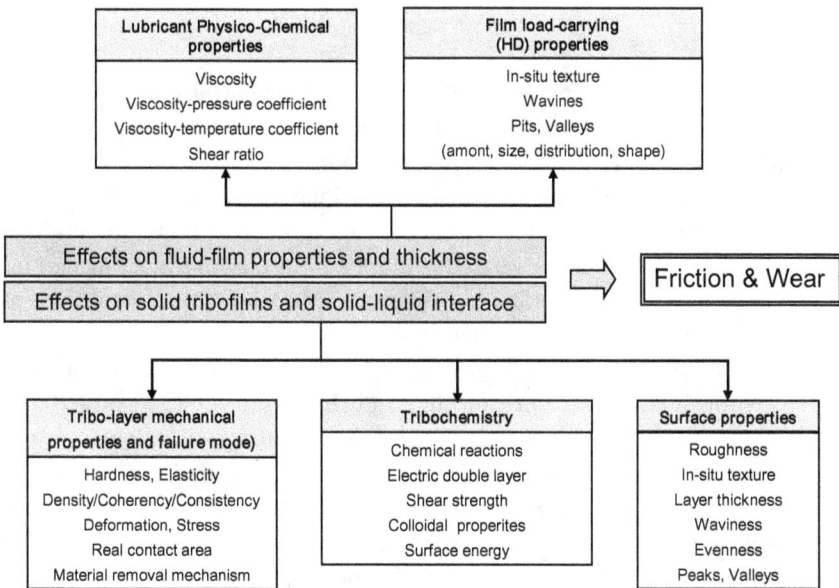

Fig. 8.26. A schematic illustrating the parameters affecting fluid-film properties, tribofilms and the solid-liquid interface, which depend on the surface charge and pH.

boundary- and/or mixed-lubrication films, and consequently the friction
and wear under such tribological conditions. This seems to be especially
important for water-based and other thin lubricating films, which due to
their low viscosity and their dependence on surface charge, provide poor
and/or fragile film conditions and a corresponding load-carrying capacity
in the tribological contacts.

Finally, another point should be mentioned; this Chapter only deals
with "pure" water lubrication, without any additive aid. However, in many
water systems, additives need to be added for various purposes: anti-
algae, anti-bacterial, temperature-stabilizers, etc. With increasing interest
in water systems, additives for improving tribological properties will also
play a more important role in the future. New additives have been developed over the last few years, that are also suitable also for ceramic aqueous
lubrication. Some very promising results for boundary lubrication of silicon
nitride and silicon carbide under aqueous conditions were recently presented
with brush-like copolymer additive, PLL-g-PEG.[56] There is no doubt that
this area will further develop, both with the need for green lubrication technology, as well as due to improvement in different areas of nanotechnology
that are required to achieve this.

8.7. Conclusions

(1) The present-day models for determining/calculating the lubricating
     film thickness and the lubrication regime typically do not consider
     effects such as surface charge, pH, zeta-potential or solubility. However,
     we have found that these electrochemical parameters affect the wear of
     alumina and zirconia by an order of magnitude, and their friction by a
     factor of 2–3. So, the estimated contact conditions and lubricating-film
     properties can be significantly different from the actual conditions if
     surface-charge effects are not properly considered.
(2) We have recognized several groups of parameters that directly influence
     the properties of the lubricant film and the severity of the contact con-
     ditions, thus the friction and wear under tribological actions, i.e. the
     surface-layer properties, the physico-chemical and load-carrying prop-
     erties of the lubricating film, the tribochemical reactions and the tribo-
     layer failure modes of water-lubricated contacts. More than 25 specific
     properties from the above groups of properties were shown (Fig. 8.26).
(3) We suggest that the surface charge and the associated pH and zeta-
     potential should be controlled and used for a better estimation of the

contact conditions in boundary-lubricated contacts. By considering the surface-charge (pH) effects, the description of the aqueous solution will improve significantly and consequently the reliability and repeatability of the experimental data and their match with the actual conditions will also improve.

(4) Due to the increased number of boundary (and mixed) lubricated systems, which are very sensitive to the contact conditions within a small variation of surface features, more tailored models that consider a more complete set of tribo-electro-chemical parameters, including the surface charge and pH effects, should be developed in the future. This seems to be especially important for water-based and other thin lubricating films, which due to their low viscosity and their dependence on surface charge, provide poor and/or fragile film conditions and a corresponding load-carrying capacity in the tribological contacts.

References

1. T. E. Tallian, *ASLE Trans* 10, 418 (1967).
2. G. W. Stachowiak and A. W. Batchelor, *Engineering Tribology Third Edition*, Elsevier Butterworth-Heinemann, Oxford, UK (2005).
3. K. Vercammen, K. Van Acker, A. Vanhulsel, J. Barriga, A. Arnšek, M. Kalin and J. Meneve, *Tribol Int* 37, 983 (2004).
4. J. Barriga, M. Kalin, K. Van Acker, K. Vercammen, A. Ortega and L. Leiaristi, *Wear* 261, 9 (2006).
5. M. Kalin, J. Vižintin, K. Vercammen, J. Barriga and A. Arnšek, *Surf Coat Technol* 200, 4515 (2006).
6. M. Kalin and J. Vižintin, *Wear* 261, 22 (2006).
7. M. Kalin, F. Majdič, J. Vižintin, J. Pezdirnik and I. Velkavrh, *J Tribol* 130, 11013 (2008).
8. F. Majdič and J. Pezdirnik, *Ind Lubr Tribol* 62, 136 (2010).
9. F. Majdič and J. Pezdirnik, *Stroj Vestn – J Mech E* 54, 841 (2008).
10. L. Jordi, C. Iliev and T. E. Fischer, *Tribol Lett* 17, 367 (2004).
11. M. Chen, K. Kato and K. Adachi, *Tribol Int* 35, 129 (2002).
12. T. E. Fischer and H. Tomizawa, *Wear* 105, 29 (1985).
13. X. Dong and S. Jahanmir, *Wear* 165, 169 (1993).
14. R. S. Gates, S. M. Hsu and E. E. Klaus, *Tribol Trans* 32, 357 (1989).
15. Y. S. Wang, S. M. Hsu and R. G. Munro, *Lubr Eng* 47, 63 (1991).
16. M. G. Gee, *Wear* 153, 201 (1992).
17. S. Jahanmir and X. Dong, *J Tribol* 114, 403 (1992).
18. M. Kalin, S. Jahanmir and G. Drazic, *J Am Ceram Soc* 88, 346 (2005).
19. M. Kalin, B. Hockey and S. Jahanmir, *J Mater Res* 18, 27 (2003).
20. H. Tomizawa and T. E. Fischer, *ASLE Trans* 30, 41 (1987).
21. S. Jahanmir and T. Fisher, *Tribol Trans* 32, 32 (1989).

22. P. Andersson, *Wear* 154, 37 (1992).
23. K. Komvopoulos and H. Li, *J Tribol* 114, 131 (1992).
24. R. H. J. Hannink, M. J. Murray and H. G. Scott, *Wear* 100, 355 (1984).
25. T. E. Fischer, M. P. Anderson, S. Jahanmir and R. Salher, *Wear* 124, 133 (1988).
26. S. Novak, G. Dražič and M. Kalin, *Wear* 259, 562 (2005).
27. J. Xu and K. Kato, *Wear* 245, 61 (2000).
28. R. S. Gates and S. M. Hsu, *Tribol Trans* 34, 417 (1991).
29. R. S. Gates and S. M. Hsu, *Tribol Letters* 17, 399 (2004).
30. M. Chen, K. Kato and K. Adachi, *Wear* 250, 246 (2001).
31. R. Iler, *The Chemistry of Silica*, Wiley, New York, pp. 625–632 (1979).
32. K. Takeuti, *Chemistry of Adsorption*, Sangyotosyo, Tokyo, pp. 42–63 (1995).
33. V. A. Muratov, T. Luangvaranunt and T. E. Fischer, *Tribol Int* 31, 601 (1998).
34. M. G. Gee and N. M. Jennett, *Wear* 193, 133 (1996).
35. B. Basu, R. G. Vitchev, J. Vleugels, J. P. Celis and O. Van Der Biest, *Acta Mater* 48, 2461 (2000).
36. M. Kalin, S. Novak and J. Vižintin, *Wear* 254, 1141 (2003).
37. S. Novak and M. Kalin, *Tribol Letters* 17, 727 (2004).
38. J. K. Lancaster, Y. A. Mashal and A. G. Atkins, *J Phys D: Appl Phys* 25, A205 (1992).
39. M. Kalin, S. Novak and J. Vizintin, *J Phys D: Appl Phys* 39, 3138 (2006).
40. M. Kalin, G. Dražič, S. Novak and J. Vižintin, *J Eur Ceram Soc* 26, 223 (2006).
41. M. Kosmulski, *J Colloid Interf Sci* 337, 439 (2009).
42. G. H. Kelsall, Y. Zhu and H. A. Spikes, *J Chem Soc Faraday Trans* 89, 267 (1993).
43. A. Marti, G. Hahner and N. D. Spencer, *Langmuir* 11, 4632 (1995).
44. G. Hahner, A. Marti and N. D. Spencer, *Tribol Lett* 3, 359 (1997).
45. D. C. Prive, S. G. Bike, *Chem Eng Commun* 155, 149 (1987).
46. W. H. Strehlow, E. L. Cook, *J Phys Chem Ref Data* 2(1) p. 169 (1973).
47. A. Feiler, I. Larson, P. Jenkins and P. Attard, *Langmuir* 16, 10269 (2000).
48. A. Kitahara and A. Watanabe (Eds), *Electrical Phenomena at Interfaces. Fundamentals, Measurements and Applications*, Marcel Dekker, Inc., New York, (1984).
49. S. P. Rao, S. S. Triopathy and A. M. Raichur, *Colloid Surface A* 302, 553 (2007).
50. R. H. J. Hannick, M. J. Murray and H. G. Scott, *Wear* 100, 355 (1984).
51. B. Basu, R. G. Vitchev, J. Vleugels, J. P. Celis and O. Van der Biest, *Acta Mater* 48, 2461 (2000).
52. J. Takadoum, *Wear* 170, 285 (1993).
53. X. Wang, K. Kato, K. Adachi and K. Aizawa, *Tribol Int* 36, 189 (2003).
54. X. Wang, K. Kato, K. Adachi and K. Aizawa, *Tribol Int* 34, 703 (2001).
55. A. Kovalchenko, O. Ajayi, A. Erdemir, G. Fenske and I. Etsion, *Tribol Int* 38, 219 (2005).
56. W. Hartung, A. Rossi, S. Lee and N. D. Spencer, *Tribol Lett* 34, 201 (2009).

Index

18-methyleicosanoic acid (18-MEA), 107

acidic, 265
acidic environments, 251
additives, 239, 266
adhesion, 7, 79, 95, 96, 248
adhesion force, 110
adsorption, 82, 97, 116
AFM, 6, 194, 248
aggrecan, 1
Al-hydroxide, 244
albumin, 15
Alexander and de Gennes, 188
alkaline conditions, 265
alkaline environments, 251
alkanethiol, 14
alumina, 237, 266
aminofunctionalized silicones, 133
amodimethicone, 133
aminothiol, 14
amylase, 96
anionic surfactants, 114
aqueous solutions, 245
articular cartilage, 1, 174
articular surface, 4
artificial lubricants, 183
asperities, 239
associated electrochemical
    phenomena, 247
atom-transfer radical polymerization
    (ATRP), 187
atomic force microscope, 196
atomic force microscopy (AFM), 6,
    106, 115
attenuated total reflection infrared
    spectroscopy (ATR-IR), 9
attractive, 248

Biesheuvel, 192
biglycan, 4

bio-surfaces, 87
biological surface, 84
biological tissues, 174
biomedical materials, 175
biosubstrates, 99
blow-drying, 124
boundary, 84
boundary film, 242
boundary lubricants, 92
boundary lubrication, 89, 145, 166
boundary- and/or mixed-lubrication
    films, 266
boundary-lubricated contacts, 267
boundary-lubrication, 237
boundary-lubrication film, 264
boundary-mixed regime, 99
brittle, 244
brittle fracture, 246
brush, 184
brush-like copolymer additive, 266
brushing, 119
BSA, 39
BSM, 35–43, 47, 49, 65, 66

canine hip joints, 10
capillary forces, 82
capsule endoscopy, 58
carbohydrates, 221
cartilage, 1, 146
cationic polyelectrolytes, 113, 130
cationic surfactants, 112, 113, 127
ceramic, 240
cervical, 63
chain scission, 212
chitosan, 49
chocolate, 84
chondrocytes, 1
chondroitin sulphate, 2, 9
chondroitinase AC, 6
cleansing base, 136

cmc, 84
coacervates, 114
coacervation, 114
coalesce, 93
coalescence, 76, 87, 88, 94, 96
coefficient of friction, 251
cold-rolling, 75
collagen, 1, 3
colloidal silica, 242
combined mechanism, 166
combing, 119
complexes, 114
compliant, 84
conditioners, 126, 129
conditioning agents, 137
conditioning polymers, 117
contact angle, 79, 80, 82
contact stresses, 263
cosmetic, 103
cracks, 246
critical micelle concentration, 80
crosslinking, 212
crystal microbalance (QCM), 115
cuticle, 105

dangling polymer chains, 170
DE, 93
de-mineralized, 260
debris layers, 263
deep zone, 3
delamination, 263
detangling, 119
dextran brush, 205
diamondlike carbon, 230
diffusion layer, 242
dimethicones, 133
displacement energy, 76, 79, 82, 83
dissipated energy, 248
distilled, 260
distilled water, 251
DLC-coated surfaces, 227
DN hydrogel, 175
DOPA, 45
double network, 146
drainage time, 157
dynamic-concentration model, 77

ECR, 224
eel-like, 170
effective contact area, 157
EHD, 77, 99
EHL, 223
EHL lubrication, 226
elasticity, 164, 262
elasto-hydrodynamic regime, 89
elastohydrodynamic, 19, 74, 87
elastohydrodynamic lubrication, 169
electric double layer, 247
electric forces, 249
electrical double layer, 83, 242
electro-chemical parameter, 251
electro-chemical properties, 264
electro-kinetic repulsive force, 248
electrochemical, 245
electrochemical mechanism, 246
electrochemical parameters, 237, 266
electrolyte, 248
electron microscopy, 4
electrostatic double layer, 162
electrostatic forces, 248
emulsifier, 75, 82
end-grafted chains, 184
environmentally friendly lubricants, 73
ESEM, 6

failure mode, 263, 266
fat, 84, 87, 89, 90, 96, 98
fiber-to-fiber, 110
fiber-to-fiber friction, 110
fiber–fiber interaction, 110
fibroblast-like synoviocytes, 21
film, 260
film thickness, 75, 77, 238, 259, 266
food, 73, 84, 85, 90, 91, 94, 96–99
friction, 2, 251
friction coefficient, 111
friction on ice, 219
friction reduction, 173
frictional forces, 248

Garcia combing, 123
gastrointestinal, 58, 59
gel, 146
gel modulus, 167
geometric effects, 169
glycerol, 223
glycoproteins, 183
good solvent, 183
grafting density, 192
grafting-from, 186
grafting-to, 186
green lubrication, 266
Grest, 195
grooming, environmental and
   chemical stresses, 108

hair, 105
hair assemblies, 110
hair care products, 126
hardness, 262
high mechanical strength, 175
high mechanical toughness, 175
high shear rates, 98
high shear viscosity, 97
human ocular mucins, 41
HWS, 55
hyaline cartilage, 1
hyaluronan, 1, 17
hyaluronidase, 9
hydrated amorphous silica, 242
hydrated layer, 244
hydrated lubrication, 166
hydrated tribochemical layers, 244
hydrocolloid, 97, 98
hydrodynamic, 92, 239
hydrogel friction and lubrication, 145
hydrogels, 146
hydrogen bonds, 242
hydroxides (Zr-OH), 244, 258
hydroxyproline, 9

IEP, 247
interface shear strength, 264
interfacial interaction, 160
interfacial shear, 264

interfacial tension, 80
interpenetration zone, 190
intima, 1
ironing, 73, 74
iso-electric-point (IEP), 135, 247
iso-viscous elastohydrodynamic
   lubrication, 85

Karl Fischer analyses, 229
keratin, 105
keratin sulphate, 9
Klein, 190

lambda, 238
lamina splendens, 5
Langmuir-Blodgett deposition, 187
Langmuir-Schaeffer, 188
lateral force, 248
latex, 15
layers, 244
LED UV source, 210
LFM, 248
LFM or FFM mode, 116
ligand-receptor, 45
link protein, 15
lipids, 9
lipophilic conditioners, 129
load-carrying capacity, 266
low friction coefficients, 171
low shear, 242
low-friction gel, 178
LUB-C, 21
lubricant, 159
lubricating film, 237
lubrication mechanism, 110
lubrication regime, 238, 259, 266
lubricin, 1, 14

macrosurfactant, 188
master curve, 92
Maxwell stress field, 248
mayonnaise, 74, 84, 95
mean-field theory, 186
mechanical properties, 261
metal sheet rolling, 73

metallic sheet rolling, 74
mica, 16, 195
micelles, 95
microemulsions, 95
micropattern, 61
middle zone, 3
Milner, 190
miniemulsions, 130
mixed, 239
mixed lubrication, 92
mixed regime, 89
mixed-flow model, 77
mixed-lubrication regime, 84, 97
molecular dynamics, 184
molecular-dynamics simulations, 231
morphology, 261
MUC5, 54
mucoadhesion, 42
mucoadhesive, 43, 59, 60
mucous, 88
mushroom, 184
myoinositol, 222

nanolubrication, 208
nanoscale friction, 248
net surface charge, 247
neutron-reflectivity measurements,
    195
NMR spectra, 229
non-ionic surfactant, 83
non-lubricated, 245
non-oxide ceramics, 244

ocular mucins, 47
opacifier, 138
oppositely charged surfactants, 114
optical interferometry, 75
oral, 87
oral processing, 73, 84, 85
oral substrates, 85
oral surfaces, 96
oral tribology, 85, 99
oral-processing, 96
osmotic brush, 184
osmotic repulsion, 161

oxidative treatments, 108
oxide ceramic, 244

parallel-plate, 97
PDMAEMA, 210
PDMS, 88, 89, 93, 94, 201
peaks, 254
pearlescent, 138
PEG-silanes, 211
PEG-thiol, 187
PGM, 35, 37, 38, 45–51, 65
pH, 237, 265, 266
phase separation, 114
phenomena, 259
phospholipids, 13, 19
photoiniferter-mediated
    polymerization (PMP), 206
Pincus regime, 185
pits, 254
plating out, 76
PLL-*g*-dextran, 205
point of zero charge, 83
poly L-lysine, 14
poly(2-(dimethylamino)ethyl
    methacrylate), 210
poly(3-sulfopropyl potassium
    methacrylate (SPMK), 213
poly(3-sulfopropyl potassium
    methacrylate
    (SPMK)-co-2-(methacryloyloxy)
    ethyltrimethyl-ammonium chloride
    (MTAC)) (poly(SPMK-co-MTAC),
    213
poly(acrylamide), 207, 213
poly(dimethylsiloxane), 201
poly(L-lysine)-*g*-poly(ethylene glycol)
    (PLL-*g*-PEG), 198
poly(methacrylic acid), 211
poly(styrene)-*b*-poly(acrylic acid), 208
poly(tert-butyl
    methacrylate)-*b*-poly(glycidyl
    methacrylate sodium sulfonate)
    (PtBMA-*b*-PGMAS), 206
poly(vinylpyrrolidone)-*b*-polystyrene,
    208

polydimethylsiloxane, 86
polyelectrolyte, 160, 184, 258, 265
polyelectrolyte brushes, 171
polyelectrolyte-surfactant systems,
coacervation, 114
polyethylene oxide, 97
polyhydric alcohols, 219
polymer brushes, 183
Polymer solution, 159
polymers, 240
polypeptides, 95
polysaccharide, 97, 98
positive and negative ions, 247
precipitates, 114
PRG, 55
PRG4, 20
properties, 245
protein, 95, 96
proteoglycan, 3, 20
proteoglycan aggregate, 1
PRP-1, 49, 55
PS-b-PAA, 189
PSM, 46

rabbit stifles, 11
Raviv, 196
Rayleigh bearing effect, 262
reaction product layer, 243
real contact area, 263
relaxer treatments, 108
repulsion and adsorption model, 147
repulsive interactions, 248
reversible addition-fragmentation
chain transfer (RAFT), 187
RGM, 35, 37
rheology, 84, 85, 90, 97, 99
rough, 99
rough substrates, 84
roughness, 85–87, 89, 90, 164, 254
running-in, 242

saliva, 33, 34, 52–58, 65, 95, 96, 99
salted-brush regime, 186
scaling laws, 184
scaling theory, 188

self-consistent mean-field theory, 184
SEM, 6
sensory, 84, 90, 91, 96, 99
SFA, 14
shampoo, 117, 126
shampoo formulation, 136
shear, 264
Si_3N_4, 241
SiC, 241
silica, 241
silica gel, 244
silicic acid, 244
silicon carbide, 237
silicon nitride, 237
silicones, 113, 133
skin, 104
sliding, 112
sliding plateau, 112
slip plane, 242
slipperiness, 84
smooth, 261
smoothening, 243
smoothness, 84
soft and wet materials, 146
soft tissues, 150
soft tribology, 84, 85
soft-tribological, 99
solubility, 237, 265, 266
spalling, 263
spreading, 95
spreading coefficient, 80, 82
squeeze-film, 19
starch, 96
starved, 76
statherin, 49, 54, 55
static peak, 112
steel, 240
Stern layer, 242
stick-slip, 111
Stribeck, 87, 90
Stribeck curve, 92, 97, 238, 243
strong-stretching theory, 184
substrates, 99
super-low friction, 241
superficial layer, 5

superficial zone, 3
superficial zone protein, 22
superlubricity, 222
surface charge, 237, 265, 266
surface chemistry, 90
surface energies, 80
surface forces apparatus, 13, 190
surface layer, 255
surface roughness, 238, 260
surface tension, 78
surface-coating, 82
surface forces apparatus (SFA), 115
surface-layer properties, 266
surfaces, 261
surfactant, 77, 78, 80, 82, 83, 89, 92, 94
swatch-on-swatch, 112
synovial fluid, 1
synovial membrane, 1
synovium, 1

ta-C, 231
tactile, 85
tangential, 3
tangential zone, 3
tap, 251
tap water, 260
taste, 85
temperature, 264
template, 170
temporomandibular joints, 10
tetragonal-to-monoclinic
    transformation, 240
texture, 254, 261
texturing, 262
thickeners, 137
thin lubricating films, 267
titanium dioxide, 248
tongue, 85–90, 95
transfer films, 239
transitional, 3
tribo-electro-chemical, 260
tribo-electro-chemical concept, 247
tribo-electro-chemical effects, 237

tribo-electro-chemical parameters, 267
tribochemical layer, 244, 246
tribochemical reactions, 237, 241, 264, 266
tribochemistry, 243
tribolayers, 252
trimethyl silylamodimethicone, 133
triple-network, 175

ultra-low friction, 174
un-evenness, 254

valleys, 254
viscosity, 76, 91–94, 97, 98, 240, 263
viscous hydrodynamics, 2
viscous-like, 244

water, 111, 240
water lubrication, 237
water solutions, 265
water-based, 267
water-lubricated conditions, 245
water-lubricating films, 264
waviness, 254, 261
wear, 2, 177, 251
wear debris, 246, 252, 256
wear mechanisms, 257, 264
wear-promoting, 259
wear-protective, 259
wear-protective tribochemical layers, 240
wear-resistance, 250
wet, 77, 93
wetted, 92
wetting, 78, 79, 82, 86, 90, 94, 95
work of adhesion, 79, 82, 93

xanthan gum, 97, 98

Young's equation, 79

zeta potential curve, 248
zeta-potential, 237, 266
zirconia, 240, 244, 266
ZP, 247, 265

www.ingramcontent.com/pod-product-compliance
Lightning Source LLC
Chambersburg PA
CBHW050547190326
41458CB00007B/1953